T0210550

SpringerBriefs in Applied Sciences and Technology

More information about this series at http://www.springer.com/series/8884

Ryszard Bartnik · Berenika Bartnik
Anna Hnydiuk-Stefan

Optimum Investment Strategy in the Power Industry

Mathematical Models

Springer

Ryszard Bartnik
Faculty of Production Engineering
    and Logistics
Opole University of Technology
Opole
Poland

and

Department of Power Engineering
    Management
Institute of Processes and Products
    Innovation
Opole
Poland

Anna Hnydiuk-Stefan
Faculty of Production Engineering and
    Logistics
Opole University of Technology
Opole
Poland

and

Department of Power Engineering
    Management
Institute of Processes and Products
    Innovation
Opole
Poland

Berenika Bartnik
Research Department
Trigon Dom Maklerski S.A.
Warsaw
Poland

ŧ

ISSN 2191-530X   ISSN 2191-5318   (electronic)
SpringerBriefs in Applied Sciences and Technology
ISBN 978-3-319-31871-4   ISBN 978-3-319-31872-1   (eBook)
DOI 10.1007/978-3-319-31872-1

Library of Congress Control Number: 2016936275

Printed on acid-free paper

This Springer imprint is published by Springer Nature
The registered company is Springer International Publishing AG Switzerland

Contents

Notation

A	Depreciation rate
a_{el}, a_{coal}, b_{CO_2}	Control variables for prices of electricity, fuel, and CO_2 emission allowances
a_{CO_2}, a_{CO}, a_{SO_2}, a_{NO_x}, a_{dust}	Control variables for specific tariffs for emission of pollutants into the environment CO_2, CO, SO_2, NO_x, and dust
ARR	Accounting Rate of Return
b	Duration of the investment
CF	Cash Flow
D	Annual operation net profit of the power station or heat and power plant
$DPBP$	Discounted Pay-Back Period
e_c, e_{el}, e_{coal}	Specific cost of heat, electricity and coal
$E_{el,A}$	Net annual electricity output from a heat and power plant or a power station
$E_{ch,A}$	Annual use of the chemical energy of the fuel heat and power plant or a power station
E_{ch}^{gas}	Chemical energy of gas combustion in the gas turbine
E_{ch}^{coal}	Chemical energy of coal combustion in the boiler
E_{el}^{GT}	Gross electrical output of the gas turbogenerator
E_{el}^{ST}	Gross electrical output of the steam turbogenerator
F	Time variable interest (financial cost) relative to the value of investment
i	Specific investment (per unit of power)
IPP	Independent Power Producer
IRR	Internal Rate of Return
J	investment expenditure
k_c	Specific cost of heat production in a heat and power plant

$k_{c,av}$	Average specific cost of heat production
k_{el}	Specific cost of electricity production in a power unit
$k_{el,oxy}$	Specific cost of electricity production in a power unit realizing oxy-fuel process
K_{CCS}	Cost associated with the transport of carbon dioxide to the destination of its storage
$k_{el,CCS}$, $k_{CO_2,CCS}$	Specific cost of carbon dioxide transport into the place of its pumping and storage expressed, respectively, per specific unit of electricity production and unit of carbon dioxide mass
K_e	Annual exploitation cost of heat and power plant and power station
K_A	Annual operating costs of a power plant and heat and power station
K_{Ac}	Annual heat production cost in a heat and power plant
K_{coal}	Cost of fuel
K_{sw}	Cost of supplementing water in system circulation
K_{sal}	Cost of remuneration including overheads
K_{serv}	Cost of maintenance and overhaul
K_m	Cost of non-energy resources and supplementary materials
K_{env}	Cost associated with the use of the environment (including charges for emission of flue gases into the atmosphere, waste disposal, waste storage, etc.)
K_{cap}	Capital cost
Kt	Taxes, local charges, and insurance
K_{CO_2}	Cost of purchasing CO_2 emission allowances
N	Calculated exploitation period of the installation
N_{el}	Gross electrical capacity of a gas or steam turbogenerator
NPV	Net Present Value
p	Charges associated with emission of harmful combustion products into the atmosphere
p	Tax rate on profit before tax
P	Income tax on gross profit
Q_A	Annual net heat output of a combined heat and power unit
Q_A	Annual net production of heat
Q_{con}	Heat from steam condensing in the condenser of the steam turbine
q_{par}	Ratio of chemical energy of gas in relation to the chemical energy of coal in the parallel system

q_{ser}	Ratio of chemical energy of gas in proportion to the chemical energy of coal in the series system
R	Loan installment
r	Discount rate
ROE	Return on Equity
ROI	Return on Investment
S_A	Annual revenues from the operation of heat and power station or from power plant
$SPBP$	Simple Pay-Back Period
t	Time
T	Exploitation period of the installation calculated in years
u	Ratio of chemical energy of fuel to its total use, for which it is not necessary to purchase CO_2 emission allowances
v_m	Relative value of heat and electricity market
$x_{sw,m,was}$	Coefficient accounting for the cost of supplementing water, auxiliary materials, waste disposal, slag storage, and waste
$x_{sal,t,ins}$	Coefficient accounting for the cost of remuneration, taxes, insurance
x_{CCS}	Proportionality coefficient
z	Ratio of freezing an investment
Z	Annual operation gross profit of the power station or heat and power plant
δ_{serv}	Rate of fixed costs relative to investment (cost of maintenance and overhaul)
ε_{el}	Internal load of power station or heat and power plant
η	Efficiency
ρ	Emission of harmful combustion products into the atmosphere
ρ	Rate of depreciation including interest
η_B	Gross boiler efficiency
η_c	Energy efficiency of heat and electricity production
η_{el}	Gross efficiency of power plant
η_{HRSG}	Gross efficiency of the heat recovery steam generator
η_{SH}	Energy efficiency of steam header used to feed steam into the turbine
η_{GT}	Gross energy efficiency of the gas turbine
η_{ST}	Energy efficiency of steam turbine
η_{me}	Electromechanical efficiency of the steam turbogenerator
σ_A	Annual cogeneration index

Chapter 1
Introduction

For the things of this world cannot be made
known without a knowledge of mathematics
(Roger Bacon, 1214–1294).

Abstract This chapter presents the methodology of seeking optimum investment strategies in business.

This monograph is dedicated to the discussion of methods used for assessing economic efficiency of economic decisions regarding any enterprise, in particular with regard to investments in the power engineering sector. A decision in engineering that is not based on economic calculation is likely to be erroneous or completely false. A technical decision, despite being important and needed, can only be applied in search of improving engineering processes and design solutions in machines and apparatuses. In market economy it is the economic criteria and income maximization that decide about the justification of a technological concept, and it is the cost-effectiveness that decides about starting of an investment. Economic criterion is superior to any technical and design aspect. However, one should emphasize here that economic analysis is only possible after a previous technical analysis. The results of engineering criteria offer input values for economic analysis.

Decisions regarding investment form some of most common long-term decisions and have a large impact on the future financial condition of enterprises whose aim is to progress through the intended production and service goals. Investments also mean financial outlay and bind resources dedicated to finance them for long, bring a result with a long time perspective and are characterized with a risk.

While an investment decision is taken, it is necessary to analyze each of the considered investment projects in turn to assess and make the right decision while accounting for various components in these projects, including risk-taking strategies and uncertainty, in particular in unstable economic conditions.

© The Author(s) 2016   1
R. Bartnik et al., *Optimum Investment Strategy in the Power Industry*,
SpringerBriefs in Applied Sciences and Technology,
DOI 10.1007/978-3-319-31872-1_1

1.1   Traditional and Discounting Techniques

The principal objective of any enterprise is to bring profit. Financial profitability is the principal criterion for assessing an investment from the investor's perspective. This means that prior to a decision about engaging on an investment, investors need to make sure that the return on an investment will be sufficiently high.

The answer to a question about the profitability of investment involves the calculation of economic efficiency measures. It is additionally, necessary to conduct a study of the sensitivity of the above measures to assess the changes in their value in the function of other parameters which influence them. The analysis of sensitivity gives a larger view in terms of profitability of an investment and enables the investor to assess its security. In the conditions of competitive economy, it gives the potential to perform a price policy.

The methods of economic effectiveness used in practice are classified into two categories:

- traditional (simple) methods of assessing efficiency, in the form of such measures as: *ROI* (*Return on Investment*), *ROE* (*Return on Equity*), *ARR* (*Accounting Rate of Return*), and *SPBP* (*Simple Pay Back Period*);
- methods involving discounted financial transfers, in which assessment parameters took the form: *NPV* (*Net Present Value*), *IRR* (*Internal Rate of Return*), *DPBP* (*Discounted Pay Back Period*).

Traditional methods are the ones not accounting for the change in the value of money in time and based on profit as the measure of the net benefit. At the same time, the methods which account for this variation and accounting for net benefits in the category of cash flows are known as *discounting methods*. The latter ones deal with the entire period in which an enterprise is in operation, i.e. include the duration of the investment and exploitation period, in which an economic gain is forecasted. In the literature of the subject they are called dynamic methods of the investment account, in contrast to more traditional ones, sometimes called static methods of the investment account. It is noteworthy that on the basis of simple rate of return it is not possible to create objective decision criteria; hence, maximization of the value and gaining a value that is higher than the boundary level form the criteria for taking a decision based on ROI and ROE ratios. The determination of a way to determine a boundary level of a rate is, however, strictly subjective. It is therefore, postulated here that the traditional methods should be only applied at the initial stages of identifying rates in the process of absolute assessment of profitability of investments. In addition, traditional measures should only be observed with regard to enterprises taken on a small scale and with respect to a relative short economic life cycle. The application of discounting methods for absolute assessment of profitability of an enterprise can aid in the decision about an effective and correct investment policy.

One needs to bear in mind that all measures of economic efficiency of an investment, i.e. both simple and discounting ones, do not account for several components connected directly with its realization and subsequent exploitation. Such elements include among others:

- influence of time,
- effect of risk.

The influence of time and the risk associated with an enterprise is hard to predict. In the conditions of uncertainty regarding the future financial condition, social tensions, changing interest rates, legal regulations, any investment in a business project is particularly difficult. In such conditions, an investor is likely to hesitate before investing financial resources even in the circumstances when the profitability of an investment could be potentially very high.

This monograph focuses on a discussion of basic issues relating to the application of discounting methods of assessing economic efficiency of an investment, i.e. ones involving the change of the value of money in time and accounting for net benefits in terms of cash flows. Discounting methods are considered correctly to be more effective criteria for decision making than traditional methods not accounting for the variation in the value of money in time. Among these, the method applying Net Present Value (NPV) is considered as the one with least disadvantages. What is more, this monograph presents an original discounting methodology of assessing the economic efficiency of an investment with a record relating to continuous time. Until this date, the measures of economic efficiency of an arbitrary enterprise are presented in the literature by means of discreet records (discount rate itself is a geometrical progression) and can be applied only in this form. From that, the total net profit is defined by the formula [1] (cf. formulae (3.1)–(3.3), Chap. 3.2):

$$NPV = \sum_{t=1}^{N} \frac{CF_{A,t,net}}{(1+r)^t} - J_0 \qquad (1.1)$$

(a search for an optimum investment strategies in business should be adopted for the target of $NPV \rightarrow$ max) and by means of it and under the assumption that $NPV = 0$, we can define IRR rate and $DPBP$ duration [cf. formulae (2.13), (2.14)]:

$$\sum_{t=1}^{N} \frac{CF_{A,t,gross}}{(1+IRR)^t} = J_0 \qquad (1.2)$$

$$\sum_{t=1}^{DPBP} \frac{CF_{A,t,net}}{(1+r)^t} = J_0 \qquad (1.3)$$

while: $CF_{A,t,net}$—annual net cash flow in the successive years $t = 1, 2, \ldots,$ N (Fig. 1.1), which is given by the difference between revenues S_A (formula 1.10) from the sales of products (e.g. electricity) and expenses (exploitation cost K_e,

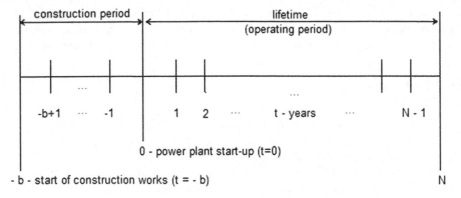

Fig. 1.1 Time line representing the course of an investment enterprise

formula (1.11), and corporate tax P (formula 1.4) paid on the annual gross profit Z (formula 1.12). The cost K_e evidently does not include depreciation since it does not occur during the exploitation; depreciation in the formulae (1.1)–(1.3) is equal to J_0);

$$CF_{A,t,net} = S_{A,t} - K_{e,t} - P_t \qquad (1.4)$$

where the rate of corporate tax is expressed by the formula [cf. formula (2.16)]:

$$P = (S_A - K_e - \rho J_0)p \qquad (1.5)$$

$CF_{A,t,gross}$—gross cash flow; this cash flow does not include corporate tax P

$$CF_{A,t,gross} = S_{A,t} - K_{e,t} \qquad (1.6)$$

where: J_0—discounted investment J associated with the development of the enterprise at the instant when the investment starts $t = 0$ (Fig. 1.1)

$$J_0 = zJ \qquad (1.7)$$

(the investment $J_0 = zJ$ is meant be regained, which means that it undergoes depreciation),

N calculation period of the enterprise expressed in years (Fig. 1.1),

r discount rate (rate of interest on investment capital which accounts for the change of the value of money in time),

t successive years of the exploitation of an enterprise, $t = 1, 2, \ldots, N$ (Fig. 1.1),

p rate of corporate tax on annual gross profit Z,

z discount coefficient (coefficient of investment freezing) on the investment capital J at the instant when the investment is complete (Fig. 1.1), $z > 1$; this coefficient includes the undesired effect on freezing capital during the investment, as this capital does not bring any profit, while the interest on the capital J increases [1]

$$z = \frac{(1+r)^{b+1} - 1}{(b+1)r} \tag{1.8}$$

b—duration of the investment in years (Fig. 1.1).

If $b = 0$, then, consequently, $z = 1$. The longer the period b, the value of z is greater from 1, and hence, the higher the value of J_0. Therefore, more expensive investment means that the lower is the total value of *NPV* (formula 1.1).

The freezing coefficient was obtained for the case when the investment J has a uniform distribution in time for the duration of the investment, $\Delta J_t = \Delta J = J/(b+1)$, and when they are additionally paid at the beginning of each year of the investment. By using the formula to express the sum $(b + 1)$ by means of the first terms of the geometrical series, a formula is derived in the form

$$J_0 = zJ = \sum_{t=-b}^{t=0} \frac{\Delta J_t}{(1+r)^t} = \sum_{t=-b}^{t=0} \Delta J_t (1+r)^{|t|} = J\frac{(1+r)^{b+1} - 1}{(b+1)r} \tag{1.9}$$

The most beneficial situation for the investor would be one in which the total investment J could be paid to the contractor after the completion of the investment instead of only its agreed $\Delta J_{t=0}$ part, i.e. at the instant marked in time as $t = 0$ (Fig. 1.1). In this case the investor does not have to bear any financial cost (i.e. interest) on the investment J. In this case, the freezing coefficient would assume its minimum value, i.e. equal to $z_{min} = 1$ and, consequently, $J_0 = J$. Most financial disadvantages are associated with a condition in which the total value of the investment J would have to be paid at the instant when it starts ($t = -b$) (Fig. 1.1). Consequently, the freezing coefficient assumes its highest value, equal to $z_{max} = (1 + r)^b$, and, hence, the investment takes on its highest cost. This cost, which is formed by the difference between the discounted investment J_0 and undiscounted one J, would be equal to $J_0 - J = J(1 + r)^b - J$. In practice, freezing coefficient should be derived on the basis of a scenario defined in the contract of investment, i.e. for the successive installments ΔJ_t defined in it, as they are paid to the investor in the successive years t of its duration. The initial calculations involving the initial assessment of economic effectiveness of an investment, called *Pre-feasibility Study*, it is possible to determine the value of the freezing coefficient with the use of the formula in (1.8).

In accordance with the definition of *IRR* rate, discounted investment $J_0 = zJ$ on the right-hand size of formula (1.2) constitutes its function as well [cf. formula (2.13)].

For the case of heat and power plant, the annual revenues S_A in the formulae (1.1)–(1.3) account for the annual revenues from the sales of heat and electricity, respectively, in the form

$$S_{A,t} = Q_{A,t}\, e_c + E_{el,A,t}^{CHP}\, e_{el}, \tag{1.10}$$

where:

e_c, e_{el}	specific prices of heat and electricity,
$Q_{A,t}$,	annual net production of heat and electricity in the heat and power
$E_{el,A,t}^{CHP}$	plant.

For the case of the power station, the revenue S_A is only gained from the sales of electricity and it is gained from the sales of electricity and hence, we have to substitute zero in the place of Q_A in the formula (1.10).

The annual operating cost K_e of the power station and heat and power plant includes: the cost of fuel K_{coal}, cost of supplementing water in system circulation K_{sw}, cost of remuneration including overheads K_{sal}, cost of maintenance and overhaul K_{serv}, cost of non-energy resources and supplementary materials K_m, cost associated with the use of the environment K_{env} (including: charges for emission of flue gases into the atmosphere, waste disposal, waste storage, etc.), taxes, local charges and insurance K_t, cost of purchasing CO_2 emission allowances K_{CO_2}:

$$K_e = K_{coal} + K_{sw} + K_{sal} + K_{serv} + K_m + K_{env} + K_t + K_{CO_2} \tag{1.11}$$

The cost K_{CO_2} directly results from the climate policy of the old 15 EU member states. This, in turn, could potentially lead to the increase of the operating cost of the power station and heat and power plant by as much as several times.

The value $S_A - K_e - \rho J_0$ in formula (1.5) represents the mean annual gross profit

$$Z = S_A - K_A = S_A - (K_e + z\rho J) \tag{1.12}$$

The annual net profit is expressed by the formula

$$D_A = Z - P = Z(1 - p) \tag{1.13}$$

The term $z\rho$ in the formula (1.12) denotes the annual pay-back period associated with the repayment of the capital investment and annual interest on it. The value ρ is the discounted annual depreciation rate ρ and is expressed by the formula [1]

$$\frac{1}{\rho} = \sum_{t=1}^{N} \frac{1}{(1+r)^t} = \frac{(1+r)^N - 1}{r(1+r)^N} \tag{1.14}$$

The sum $K_e + z\rho J$ in the formula (1.12) denotes the total annual cost K_A associated with the operation of the power station or the heat and power plant, total annual operation cost K_e and the capital cost (depreciation including interest, i.e. annual capital instalment including interest) $K_{cap} = z\rho J$ associated with the production of heat and electricity

$$K_A = K_e + z\rho J \tag{1.15}$$

The difference between the annual revenues and expenses $CF_A = S_A - K_e$ (also known as the *annual operating profit*) offers the possibility to at least balance the cost of depreciation (*depreciation*, as mentioned does not constitute an expense but forms a deduction of financial resources from the flow CF_A in order to regain the expenditure made for investment) and the interest on the investment capital $z\rho J$. The investment is more profitable to the same extent to which the annual profit Z is greater. For the case when $Z = 0$, then after N years of exploitation of the power station or heat and power plant only the investment will be regained with the interest calculated on the basis of the discount rate for the entire calculation period (i.e. for the period including investment and operation of the power station and heat and power plant), whose sum is equal to $b + N$ (Fig. 1.1). For the case when $Z < 0$, the investment will bring a loss.

For the case of heat and power plant, the formula (1.12) takes the form

$$Z = Q_A(e_c - k_c) \tag{1.16}$$

while for the power station it is in the form

$$Z = E_{el,A}(e_{el} - k_{el}) \tag{1.17}$$

where: $E_{el,A}$—annual net electricity production in the power station, while the specific cost of heat production in the heat and power plant is equal to [cf. formula (2.54)]

$$k_c = \frac{K_e + Jz\rho - E_{el,A}^{CHP} e_{el}}{Q_A} \tag{1.18}$$

and the specific cost of electricity production in the power station is expressed by the formula [cf. formula (2.43)]

$$k_{el} = \frac{K_e + Jz\rho}{E_{el,A}} \tag{1.19}$$

As mentioned above, the sum $K_e + z\rho J$ in the formulae (1.18) and (1.19) denotes the total annual cost K_A associated with the operation of the power station and heat and power plant. The numerator in formula (1.18) denotes the annual cost K_{Ac} associated with the production of the heat and power plant

$$K_{Ac} = K_e + Jz\rho - E_{el,A}^{CHP} e_{el} \tag{1.20}$$

while the revenues from the sales of the electricity in it have a negative sign $-E_{el,A}^{CHP} e_{el}$, as they form the loss associated with lack of heat production.

The discreet notations (1.1)–(1.3) have a number of considerable disadvantages, which are avoided by using notations in continuous time (formulae (2.12)–(2.14), Chap. 2.2). Moreover, the notations in continuous time using NPV, IRR and DPBP parameters and the development of continuous mathematical models with the application of them for the purposes of analysis of economic efficiency of investments enable the application of differential calculus for the study of their courses in time. This, in turn, provides other important information, which would otherwise be very difficult or impossible to observe.

Reference

1. Bartnik, R., Bartnik, B.: Economic calculus in power industry (Wydawnictwo Naukowo-Techniczne WNT), Warszawa (2014)

Chapter 2
A Formulate of Problem of Seeking an Optimum Investment Strategy in Power Engineering

Abstract The applied energy technology and its technical solution is decisive for the value of an investment (J_0) for building an energy source (generally it can be a source of electric energy and heat). So it determines the amount of financial costs (F) and loan installment (R) in its annual activity costs in successive years (t = 1, 2, ...N) together with energy carriers prices and specific charges for emitting pollutants to the environment it also determines annual revenues (S_A) and yearly operating costs i.e. the Net Present Value (NPV). Thus the optimum investment strategy for selecting the technology will be this one for which the calculated NPV value, with the application of Bellman's principle of optimality and in particular Pontryagin's maximum principle, reaches its maximum with the assumed value of N_{el} electrical capacity of a power station. Attention should be paid to the fact that all investment decisions are the long-term ones so integrally connected with the risk of failure. Influence of time and the risk connected is very difficult and, as it is mentioned before, almost impossible to predict—especially in unstable economic conditions. But this instability does not release any investor from the duty to search for an optimum investment strategy as this search enables the investor to undertake an analysis on the basis of scientific forecasts. It allows thinking about it in a scientific manner and an analysis of conditions like changes in price relations between energy carriers or costs of utilizing the environment and so on, with which the strategy should be changed. So, the maximum functionality search results show how the mentioned price relations and environmental tariffs influence the optimum investment strategy i.e. the selection of an optimum energy technology. Besides, one of the methods of minimizing the risk can involve diversification of applied technologies. This means that it is necessary to discuss available technology options. As a consequence, one can rationally diversify the processes so as to make a choice among the most economically effective ones. Also, what is very important, the safety of electricity supplies will grow. So the application of mathematical models in economy and their analysis with the use of assumed scenarios allows a rational selection of investment strategies enabling achievement in the nearest future of the desirable values in an optimum way.

© The Author(s) 2016

R. Bartnik et al., *Optimum Investment Strategy in the Power Industry*,
SpringerBriefs in Applied Sciences and Technology,
DOI 10.1007/978-3-319-31872-1_2

Investing funds in any business venture should be preceded by finding the optimal investment strategy. For example, investments in energy brings a need to search for an answer to the following questions:

- What technologies should be applied for this purpose?
- What effect on the final value of the adopted target criterion do the prices of energy carriers have and what are the relations between them?
- How is it possible to phase the resources obtained from external sources over the years in order to achieve the assumed target in an adopted horizon?

The following questions focus on the economic effectiveness of investment in the power sector. It is quite obvious that the economic result should be as high as possible while the costs associated with the generation of the electricity should be as low as possible.

In order to analyze a given technical, economic phenomena or technical-economic ones, it is necessary to develop a mathematic model of it, i.e. establish a mathematical notation of the phenomenon which describes its function space. It is an all-time truth which states: *For the things of this world cannot be made known without a knowledge of mathematics*, as expressed in 13th century by Roger Bacon (1214–1294).

We sometimes have to do with an opinion stated by economists that mathematical models which describe phenomena in economics consequently give it some weaknesses, or are totally unnecessary and even harmful. This claim is based on the statement that economics is a social and arts discipline. One cannot agree with any view of this kind. If a single mathematical model leads to errors and economic failure, this does not mean that modeling is generally wrong. It only indicates that a specific model was of poor quality. It also indicates that wrong assumptions were taken into account and a faulty theory which was taken as the basis for a model. The above principle is described by the GIGO (Garbage In, Garbage Out) phrase which says that if most nonsensical data (garbage in) is input into a system, the data will be produce nonsensical output (garbage out). If calculations using a faulty model apply even correct input data, the results from the calculations will be wrong as well. In the reverse case, if a model is correct and the data input into is nonsensical, the results will be wrong as well. In addition, if a model will be garbage quality, the output will be garbage as well.

What is worse when politicians will enter this into force, it may lead to crises and may finally end up with years of economic collapse on global scale if such a politicians will be a leaders of world powers.

It is also possible to encounter a case when a phenomenon or variable is not taken into account while it should have been considered in a given model. Therefore, one needs to know that the final effect of any model cannot yield more information that is determined by the variables inherent as part of the modeled issue. Moreover, computer simulations gained a result of such modeling do not consider interactions caused by the unmodeled parameter. The function space will therefore not be true. For this reason it is necessary to bear in mind at all times that a model operates in a virtual space but does not analyze reality which is very complex and simply impossible to model by mathematical terms but it is only capable of analyzing a set of given assumptions which need to be constantly verified. In addition, it is necessary to continuously discuss and analyze the results of the performed calculations. It is also necessary to analyze the effect of the specific parameters on the final value of the analyzed one and how sensitive the latter are to the changing value of these parameters. It involves checking whether a small variation in the input in one parameter does not cause a considerable effect on the final calculations. As a result of these considerations, one can never speak of any final results provided by a mathematical model. One needs to bear in mind that one can only determine the probability of the occurrence of a modeled phenomenon and probable results based on this phenomenon.

A couple of words need to be added regarding the stance which takes economics as a social science, i.e. economics as a branch of humanism. Such an approach equals economics as a voluble discipline, whereas important on one hand, like political sciences but leading only to bankruptcy and poverty and, hence, to destabilization and world destruction and even wars. One of the worst indication of treating economics as a social science, which is the role of demagogues and achieving particular benefits is associated with the will to print empty money, that is, money which does not have a cover in actual production. This is a way in which an economic catastrophe is delayed in time along with its consequences. In this manner a greater and greater sum of money is printed, while the effects are only faced by the citizens. The savings of the latter resulting from the long years of hard work are thus becoming worthless but only a part of trash money. As a consequence of this, prices of goods soar (one can note at this point that inflation is not associated with the increase of price but excess of money in the market whose result is visible in the rise of prices).

"Deterioration" of money always leads to disturbance and disintegration of societies and collapse of states, as the Roman Empire collapsed.

One can additionally remark that in the economic life and beyond it, it is necessary to act responsibly and fairly but never be led by populism. One can never make a list of empty promise which give unattainable hopes. A tool for this purpose, except for *Decalogue* is offered by the development of mathematical models since *For the things of this world cannot be made known…* If these models are input the functions, in which one of the independent variables is time, then such models have the capability of predicting the future and results of calculations based on them will allow one to act rationally and responsibly. Obviously, models are not capable of mirroring the reality nor the future, which is unknown (which is lucky since otherwise the world would go crazy and what would we do then?) but they allow us to perceive the inherent complex relations, which would otherwise be difficult or impossible to follow. In order to establish a correct mathematical model, it is necessary to form insightful and broad knowledge of the mind, which is based on history and psychology.

This is so as a model should involve the sense of the profit thinking which is inherent in each human. It is natural for humans to have a enjoy a state of prosperity. This spirit has been living with us for centuries and does not change. It forms the basic motivation of the human action and is totally right. Only the scenery in which humans live change with time. This spirit has formed the most powerful force guiding progress and development as the prospect of prosperity forms the strongest drive in the search for new solutions and technologies developed by humans. This sense by its very nature occurs in mathematical models which are based on an economic criterion expressed by in the idea of the potential of gaining profits. It is a superior criterion and model in the technical solutions which are strived for. Technical analysis, though important and necessary is only capable of providing ways of improving technical and technology solutions and improving engineering solutions found in machines and facilities. However, finally, it is the economic criterion, one that is based on the maximization of profit, that is decisive in the justification and selection of a specific technical solution and decides about the particular actions which are taken on by humans.

In order to develop a correct mathematical model which accounts for the above described properties and characteristics, and one which is capable of predicting the future and, thus, enables one to follow the complex relations which occur as a result of business activities taken on by humans, it is necessary to have a genius or at least wisdom worth the Nobel Price. It is one of the most basic conditions in which correct results can be obtained from a model of mathematically modeled phenomenon. It is necessary to additionally have a sophisticated mathematical apparatus, which makes it possible to research and analyze a modeled function space. In this respect, it is particularly valuable to apply the theory of optimal strategies (systems) and in particular Pontryagin's maximum principle and Bellman principle of optimality [1]. The first one find sits application in continuous processes while the latter to discreet ones.

2.1   Methodology of Seeking Optimum Investment Strategies in Business

Functional analysis, i.e. a branch of mathematical analysis which deals with the study of characteristics of functional spaces, is the domain of research into mathematical models. The term *functional* denotes a function whose argument is a function and not a number. A specific branch of functional analysis is called calculus of variations, which deals with the search for extrema of functionals, which play an important role in the study of mathematical models.

A special case of finding extrema of functionals (i.e. *Mayer problem*) is the *Lagrange's problem*, which involves the search for extrema (maxima and minima) of integral functionals:

$$J = \int_{t_O}^{t_F} F[x_1(t), x_2(t), \ldots, x_n(t); u_1(t), u_2(t), \ldots, u_n(t); t] dt = \text{ekstremum} \quad (2.1)$$

where:

$x_i(t)$ dependent variables, i.e. state variables $(i = 1, 2, ..., n)$; state variables are
coordinates of the state vector $x(t) = [x_1(t), x_2(t), ..., x_n(t)]$,

t independent variable (e.g. time), $t \in \langle t_0, t_F \rangle$,

whereas:

$$\frac{dx_i}{dt} \equiv u_i(t) \tag{2.2}$$

Lagrange's problem itself is a special case of the problem of an *optimal solution* which involves the determination of r of variable in the control $u_k = u_k(t)$ $(k = 1, 2, ..., r)$ which give extrema of an integral functional (target criterion):

$$J = \int_{t_O}^{t_F} F[x_1(t), x_2(t), ..., x_n(t); u_1(t), u_2(t), ..., u_r(t); t] dt = \text{ekstremum} \tag{2.3}$$

while the derivatives of function $x_i(t)$ fulfill in this case n number of differential equations of the first order called the state equations:

$$\frac{dx_i}{dt} = f_i[x_1(t), x_2(t), ..., x_n(t); u_1(t), u_2(t), ..., u_r(t); t] \tag{2.4}$$

while the controls $u_k(t)$ form the coordinates of the control vector $u(t) = [u_1(t), u_2(t), ..., u_r(t)]$.

The differential equations in (2.4) describe the changes of process occurring in time described by means of a functional (2.3).

The optimal strategies, i.e. ones that give the extrema of a functional (2.3) the functions of control $u_k(t)$, determine the optimal trajectory $x_i = x_i(t)$ in an n dimensional state space. In practice there is an extensive class of technical and economic solutions, in which in the place of a functional with a continuous time (2.3) and whose evolution is defined by differential equations (2.4), we have to do with processes which are discreet by their very nature. This class predominantly includes multistep tasks aimed at decision making. Economic processes are of this kind as it is always described by differential equations. The discretization step is determined by its cycle. In practice this is usually one year $\Delta t = 1$. In this case the search for an extremum of a target functional (target criteria) with the number of steps in a process to be included equal to N

$$J = \sum_{t=1}^{N} F[x_1(t), x_2(t), ..., x_n(t); u_1(t), u_2(t), ..., u_r(t); t] \tag{2.5}$$

$$= \text{ekstremum}, \quad (N = 1, 2, ...)$$

with differential state equations

$$x_i(t-1) = f_i[x_1(t), x_2(t), \ldots, x_n(t); u_1(t), u_2(t), \ldots, u_r(t); t],$$
$$(i = 1, 2, \ldots, n) \tag{2.6}$$

it is possible to apply the *Bellman principle of optimality*. This principle states that fact that all parts of an optimum trajectory optimizes a functional for the respective starting and final points. In other words, in order to ensure for a trajectory to be optimal, each of its parts (each step) needs to be optimal regardless of its initial points. This also means that the search for an optimum control needs to be performed for every step $t = 1, \ldots, N$ separately, with a respective initial point extremum value resulting from the preceding step, while each of the extrema has to be determined in accordance with the bond occurring in it. Thus, the Bellman principle allows one to search for an optimum of a functional by means of analyzing the extrema of a function and leads to a recurrent formula which expresses the total of N *Bellman equations*, in which letter S denotes the extrema of functions for steps "$t - 1$" and "t"

$$S[x_1(t), x_2(t), \ldots, x_n(t); t] = F[x_1(t), x_2(t), \ldots, x_n(t); u_1(t), u_2(t), \ldots, u_r(t); t]$$
$$+ S[x_1(t-1), x_2(t-1), \ldots, x_n(t-1); t-1],$$
$$t = 1, \ldots, N \tag{2.7}$$

For a uniform notation of formula (2.7) it was assumed that $S[x_1(0), x_2(0), \ldots, x_n(0); 0] = 0$. Functions S in (2.7) are derived by means of applying optimum controls $u_k = u_k(t)$ ($k = 1, 2, \ldots, r$) calculated from the system of equations in (2.10).

The determination of a function Eq. (2.7) with respect to function $S[x_1(t), x_2(t), \ldots, x_n(t); t]$ is equivalent to the stepwise construction of a class of optimum strategies for a variety of initial states. This task for the number of relative variables equal to $x_i(t)$ above two becomes very extensive. Such lengths lead to the use of approximated methods when solutions of specific problems are sought [1]. For less extensive problems the solution can be found more efficiently and faster by means of a method of approximations, even in the circumstances when high precision of calculations is required. In this case, the effective means involves the replacement of Bellman methodology (2.7) and differential state equations (2.6) with a continuous functional (2.3) and differential equations (2.4) and subsequent solution by *Ritz method* of approximation familiar from the calculus of variations [1].

Equations (2.6) and (2.7) present a normal (forward) recurrence. When in the search of an optimum trajectory the target is to reach the final point N with the value $x_i = x_i(N)$, Eqs. (2.6) and (2.7) needs to be substituted by backward recurrence:

$$x_i(t+1) = f_i[x_1(t), x_2(t), \ldots, x_n(t); u_1(t), u_2(t), \ldots, u_r(t); t] \qquad (2.8)$$

$$S[x_1(t), x_2(t), \ldots, x_n(t); t] = F[x_1(t), x_2(t), \ldots, x_n(t); u_1(t), u_2(t), \ldots, u_r(t); t]$$
$$+ S[x_1(t+1), x_2(t+1), \ldots, x_n(t+1); t+1],$$
$$t = N - 1, \ldots, 0 \qquad (2.9)$$

The backward recurrence, which by its nature is concerned with the backward logic of reasoning is a way of analyzing the future. It creates scientific thinking in the direction of the future. It starts with an adoption of a desired value in a year N and, subsequently, step by step, aims to model in the backward direction in order to obtain a present value which should ensure that the desired value can be achieved in year N. This backward thinking follows an optimal trajectory for an adoption target criterion and various alternatives of a control scenario i.e. time variable prices of energy carriers, specific rates associated with emission of pollutants into the environment, etc.

The calculated present value therefore indicates what technologies need to be adopted right now and imposes specific solutions needed in order to ensure that an optimum trajectory can lead to the desired target in the future. Therefore, it enables the analysis and selection of alternative solutions along with their inherent determinants in order to achieve a desired target value within a time horizon of N years. Therefore, it enables an analysis of variety of investment alternatives and technologies to be undertaken for a number of scenarios involving energy carrier prices and environmental considerations.

A noted above, the search for an optimum trajectory for a given target criterion (2.5) is associated with finding a solution (for each of successive steps t) of a system of r function equations in order to derive the r number of controls $u_k = u_k(t)$ ($k = 1, 2, \ldots, r$) which give the extremum in a given step t for a functional (2.5). This system can be presented in the following form:

$$\begin{cases} \dfrac{\partial\{F[x_1(t), x_2(t), \ldots, x_n(t); u_1(t), u_2(t), \ldots, u_r(t); t] + S[x_1(t\mp1), x_2(t\mp1), \ldots, x_n(t\mp1); t\mp1]\}}{\partial u_1} = 0 \\[2mm] \dfrac{\partial\{F[x_1(t), x_2(t), \ldots, x_n(t); u_1(t), u_2(t), \ldots, u_r(t); t] + S[x_1(t\mp1), x_2(t\mp1), \ldots, x_n(t\mp1); t\mp1]\}}{\partial u_2} = 0 \\[2mm] \qquad\qquad\qquad\cdots \\[1mm] \dfrac{\partial\{F[x_1(t), x_2(t), \ldots, x_n(t); u_1(t), u_2(t), \ldots, u_r(t); t] + S[x_1(t\mp1), x_2(t\mp1), \ldots, x_n(t\mp1); t\mp1]\}}{\partial u_r} = 0 \end{cases} \qquad (2.10)$$

in which the extremum of the function S has been determined in the preceding step $t \mp 1$ (for the case of forward recurrence for the step $t - 1$, while in backward recurrence for the step $t + 1$). In the system (2.10) prior to performing the operation of calculating partial differentials $\partial(F + S)/\partial u_k$ in the place of $x_i(t \mp 1)$ in the function S it is necessary to insert differential equations for the case of forward recurrence (2.6), and for the case of backward recurrence equations (2.8).

2.2   Target Functional in Continuous Time in Search for an Optimum Investment Strategy

During the search for an optimum investment strategy, the target criterion (optimality criterion) should involve the maximization of Net Present Value *NPV*, which is expressed by the formula in (1.1). This formula could be also presented in the form [2].

$$NPV = \sum_{t=1}^{N} \frac{S_{A,t} - K_{e,t} - F_t - R_t - \left(S_{A,t} - K_{e,t} - F_t - A_t\right)p}{(1+r)^t} \qquad (2.11)$$

which in continuous notation is expressed by the formula

$$NPV = \int_{0}^{T} [S_A - K_e - F - R - (S_A - K_e - F - A)p]e^{-rt}dt \qquad (2.12)$$

where:
A   depreciation rate,
F   time variable interest (financial cost) relative to investment J_0; interest is denoted by *F* and is an unknown function of the time variable installments *R*; $F = F[R(t)]$,
K_e variable in time exploitation costs,
N   calculated exploitation time of an enterprise expressed in years,
p   variable in time rate of income tax,
R   variable in time loan installment,
r   variable in time discount rate,
S_A variable in time annual turnover (formula 1.10),
t   time,
T   exploitation period of the installation calculated in years

   Under the condition NPV = 0 the successive measures of economic efficiency of investments can be calculated in the continuous time on the basis of formula (2.12), including Internal Rate of Return (*IRR*), value of the investment *j* and Discounted Pay-Back Period (*DPBP*) described in years [cf. formulae (1.2) and (1.3)]

$$\int_{0}^{T} (S_A - K_e)e^{-IRRt}dt = \int_{0}^{T} [F(IRR) + R(IRR)]e^{-IRRt}dt \qquad (2.13)$$

$$\int_{0}^{DPBP} [S_A - K_e - (S_A - K_e - F - A)p]e^{-rt}dt = \int_{0}^{T} (F+R)e^{-rt}dt \qquad (2.14)$$

IRR measure, defined in [2], is determined on the assumption that P income tax on gross profit Z [cf. formula (1.12)]

$$Z = S_A - K_e - F - A \tag{2.15}$$

equals to zero [cf. formula (1.5)]:

$$P = (S_A - K_e - F - A)p = 0 \tag{2.16}$$

The terms $F(IRR)$ and $R(IRR)$ on the right-hand side of Eq. (2.13) mean that the financial cost F together with the loan installment R constitute functions of the *IRR* rate, while the measures on the left-hand side of this equation constitute functions of the rate r [along with depreciation rate in formulae (2.12) and (2.14)]. The right hand sides of formulae (2.13) and (2.14) represent the discounted investment J_0 [cf. formulae (1.1)–(1.3)].

For all integrated functions in formulae (2.12)–(2.16), any change in value of a variable can be assumed, for instance we can consider any scenario involving fluctuations in the price of energy carriers in time or change in the specific tariffs for emission of pollutants into the environment. The statement of an optimality criterion (2.12) in continuous time has a considerable advantage over discrete notations. It offers the possibility of assessing the variability of the *NPV* value in order to determine its highest value while accounting for an arbitrary change in time. Moreover, such a notation enables a study of the variability in the value of *NPV* in time, which could otherwise be difficult to note, and development of its course with the application of differential calculus. Consequently, it offers the possibility of assessing the impact of a specific input value on the final result, in addition to which, it can provide a solution that is optimal along with a scope of other solutions close to such an optimum. Besides, it is possible to make remarks about the nature of changes which occur and discuss and analyse the results. It plays an important role in engineering and economics as well as in practical applications. What is more, mathematical models developed in continuous time offer conclusions to be mad about the general characteristics of a process and indicate that the course of reasoning from the general to specific details gives a potential for a reflection about the characteristics of a process. In contrast, the transfer of reasoning from the specific to the general tends to be very commonly—not to say usually—incorrect.

A selection of an investment strategy should be adopted with the goal of

$$NPV \rightarrow \max \tag{2.17}$$

for an adopted value of electrical capacity N_{el} of the power station. The values which are optimized (variables which guide decision making) include:

- available technologies and the techniques, and processes applied in the accomplishment of objectives in this technology

- its technical solutions, including the use of specific equipment, its engineering parameters, rated capacities, structure of connections, exploitation parameters of the process, etc.

In the general case, identification of an extremum of the functional integral represented in (2.11) or (2.12) along with the state equations [e.g. formula (2.20)] and constraint equations [e.g. formula (2.18)] involves finding integrated functions applied to establish the extremum of this functional. For these purposes it is necessary to apply the *Bellman's optimality principle* or *Pontryagin's maximum principle* [1, 2]. The former finds its application with regard to discreet processes (2.11) while the latter with regard to continuous processes (2.12). For the case when the forms of integrated functions are adopted in advance (i.e. by application of direct method of solving variation problems, e.g. by the *Ritz method* [1]) so as to ensure that they fulfil the boundary conditions, the task becomes a trivial one. After differentiation of the relation (2.12) within the given boundaries, finding an extremum comes down to the determination of constant values (also known as controls [1, 2]), which occur in the adopted integrated functions [in this study these include: $a_{el}, a_{coal}, a_{CO_2}, a_{CO}, a_{SO_2}, a_{NO_x}, a_{dust}, b_{CO_2}$, formulae (2.25), (2.28), (2.30)–(2.34) and (2.36)]. For this purpose, it is necessary to verify the necessary conditions for the existence of an extremum [i.e. zeroing of derivatives gained as a result of integrating functions with regard to these constant values—formula (2.40)]. The number of the necessary conditions can be as high as the number of constant values in the adopted integral functions. In other words, these conditions form a system of n equations, where n represents the number of constant values.

The continuous notation (2.12) of the optimality criterion has a considerable advantage over the discreet notation (2.11). It gives the opportunity to quickly and easily analyze the variability of the *NPV* profit so that its maximum value can be established. A singular operation of integration of the relation in (2.12) gives a compact form of the *NPV* formula (2.40), which is suitable for such an analysis, while the time-consuming and extensive process of calculating the value of the functional (2.11) step-by-step over successive years and its subsequent summation does not offer such a possibility. However, to be able to integrate the functional (2.12), all integrated functions such as the revenue S_A, operating cost K_e, financial cost F, loan installment R and depreciation installment A need to be known functions of time t. In the opposite case, i.e. when the integrated expressions are unknown functions, the identification of the extreme of the functionals in (2.11) or (2.12) imposes the need to apply Bellman's optimality principle or Pontryagin's maximum principle.

The value of the adopted electrical capacity N_{el} of the power station in optimization calculations using the *Bellman's principle of optimality* remains constant at all times. The investment associated with building a power plant involves a long-year process which consists of multiple tasks: obtaining building permission, gaining the sources of funding of the necessary investment, design stage process and construction period; therefore, the electrical capacity N_{el} has to be adopted at a constant level in search of functional maximums (2.11) and (2.12) over the entire calculation period $t \in \langle 0, T \rangle$, i.e. over the total number of years t = 1, 2, ..., N. The time variables in this period include only the control variables of state, i.e. revenues and annual operating costs of the power plant (exploitation cost plus capital costs) which additionally depend on the adopted technology and projected electrical capacity N_{el}.

The total loan installment R is imposed with a restriction. It has to be equal to the investment expenditure J_0. The relation of constraints is therefore expressed in the form:

$$\int_0^T R dt = J_0 \qquad (2.18)$$

In practice, the loan repayment installment is constant R = const and from the formula (2.18) it is obtained

$$R = \frac{J_0}{T} \qquad (2.19)$$

The depreciation rate A is expressed by the same model as loan installment R [2]. Investment expenditure J_0 is relative to the adopted technology in power industry and technical solutions and to the electrical capacity of the power plant N_{el} [2]. Therefore, it assumes different value for the same value of electrical capacity depending on the adopted technology.

In a general case, the evolution of the unknown function of the financial cost $F = F[R(t)]$ is defined by the state equation [2]

$$\frac{dF}{dt} = -rR \qquad (2.20)$$

where R is the control variable; for instance in the differential notation the above state equation for the back recurrence is expressed by the relation [2]

$$F_{t+1} = F_t - rR_t \qquad (2.21)$$

In practice, however, loan installments have a constant value R = const and the interest F is expressed by the function [2]

$$F(t) = r[J_0 - (t-1)R] \qquad (2.22)$$

In search of the maximum of the target functional (2.12), we can analyze the effect of the efficiency of the elements of equipment applied in specific technologies on its value. In this case, it is necessary to express the value $E_{ch,A}$ by means of state equations by application of energy balance equations, Figs. 2.1, 2.2, 2.3 and 2.4. For instance by application of energy balance for dual-fuel, coal-gas power station

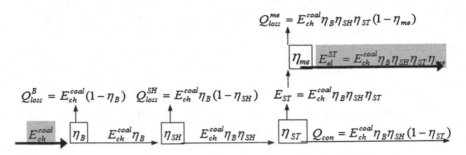

Fig. 2.1 Energy balance equation for the unit with supercritical steam parameters

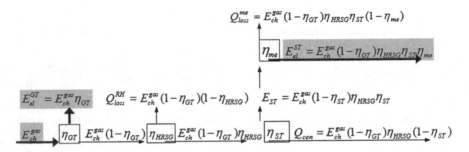

Fig. 2.2 Energy balance equation for the gas and steam power unit

operating in a parallel system (Fig. 2.3), the annual use of the chemical energy of coal in it can be expressed by the formula

$$E_{ch,A}^{coal} = \frac{N_{el}^{ST} t_A}{[q_{par}(1 - \eta_{GT})\eta_{HRSG} + \eta_B]\eta_{SH}\eta_{ST}\eta_{me}} \tag{2.23}$$

where N_{el}^{ST} is the capacity of the steam turbogenerator. It is then necessary to account for the annual use of the chemical energy of the fuel in the gas turbogenerator with a capacity of N_{el}^{GT} and efficiency η_{GT} equal to $E_{ch,A}^{g} = (N_{el}^{GT} t_A)/\eta_{GT}$, and account for the production of electricity in it, as written in Eq. (2.38). In search of the maximum of the target functional (2.12), the optimization should involve the proportion of the chemical energy of gas in relation to the chemical energy of coal combustion in the power station $q_{par} = E_{ch,A}^{g}/E_{ch,A}^{coal}$, which occurs in Eq. (2.23).

where:

E_{ch}^{gas}	chemical energy of gas combustion in the gas turbine,
E_{ch}^{coal}	chemical energy of coal combustion in the boiler,
E_{el}^{GT}	gross electrical output of the gas turbogenerator,
E_{el}^{ST}	gross electrical output of the steam turbogenerator,
Q_{con}	heat from steam condensing in the condenser of the steam turbine,

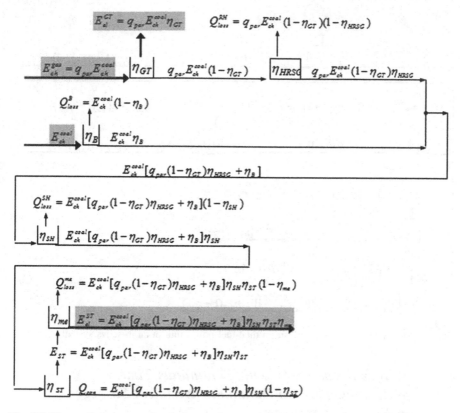

Fig. 2.3 Energy balance equation for gas and steam dual-fuel power unit in a parallel system

q_{par}	ratio of chemical energy of gas in relation to the chemical energy of coal in the parallel system,
q_{ser}	ratio of chemical energy of gas in proportion to the chemical energy of coal in the series system,
η_B	gross energy efficiency of the boiler,
η_{HRSG}	gross energy efficiency of the heat recovery steam generator,
η_{SH}	energy efficiency of steam header used to feed steam into the turbine,
η_{GT}	gross energy efficiency of the gas turbine,
$\eta_{ST} = \eta_{CR}\eta_i$	energy efficiency of steam turbine (ratio of its energy efficiency in the Clausius-Rankine cycle for condensing work and internal efficiency of the steam turbine),
$\eta_{me} = \eta_m\eta_G$	electromechanical efficiency of the steam turbogenerator (product of mechanical efficiency of the steam turbine and total efficiency of the electric generator).

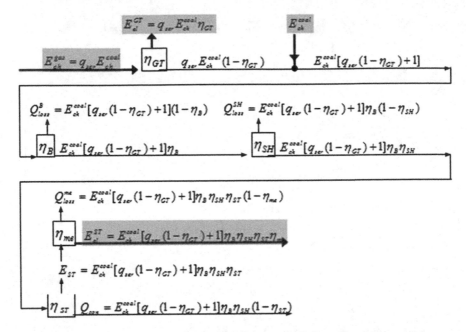

Fig. 2.4 Energy balance of gas and steam dual-fuel power unit in an in-series system

2.2.1 Mathematical Model with Continuous Time of Searching for an Optimum Investment Strategy in Electricity Sources

The remaining functions occurring in the functional (2.12) are presented below.

- *function accounting for revenues*

$$S_A(t) = E_{el,A} e_{el}(t) \tag{2.24}$$

gained from the sales of electricity, while the change of the specific price of electricity in time e_{el} (per unit of energy) can be represented by means of an exponential function (depending on the value a_{el} the price e_{el} in successive hears can increase, decrease or remain constant)

$$e_{el}(t) = e_{el}^{t=0} e^{a_{el}t} \tag{2.25}$$

The time variable measures associated with exploitation costs K_e include: the cost of fuel K_{coal}, cost of supplementing water in system circulation K_{sw}, cost of remuneration including overheads K_{sal}, cost of maintenance and overhaul K_{serv}, cost of non-energy resources and supplementary materials K_m, cost associated with the use of the environment K_{env} (including: charges for emission of flue gases into the atmosphere, waste disposal, waste storage, etc.), taxes, local charges and insurance K_t, cost of purchasing CO_2 emission allowances K_{CO_2}:

$$K_e = K_{coal} + K_{sw} + K_{sal} + K_{serv} + K_m + K_{env} + K_t + K_{CO_2} \qquad (2.26)$$

The cost K_{CO_2}, which results from the climate policy implemented by the old 15 members of the EU leads to an increase in the exploitation cost K_e of the power station. The total of the cost $K_{sw} + K_m$ and $K_{sal} + K_t$ in formula (2.26) can be accounted for as a result of a subsequent increasing the cost K_{coal} by several percent and the cost of maintenance and overhaul K_{serv} by a dozen or so percent.

- function accounting for the cost of fuel

$$K_{coal}(t) = E_{ch,A} e_{coal}(t) \qquad (2.27)$$

while the fluctuation in time of the specific price of fuel e_{coal} (per unit of energy) can be written in the form of an equation:

$$e_{coal}(t) = e_{coal}^{t=0} e^{a_{coal} t} \qquad (2.28)$$

- function accounting for the use of natural environment

$$
\begin{aligned}
K_{env}(t) = E_{ch,A} \big[& \rho_{CO_2} p_{CO_2}(t) + \rho_{CO} p_{CO}(t) \\
& + \rho_{NO_x} p_{NO_x}(t) + \rho_{SO_2} p_{SO_2}(t) + \rho_{dust} p_{dust}(t) \big]
\end{aligned}
\qquad (2.29)
$$

while the fluctuation in time of the specific price of emission charges of CO_2, CO, NO_x, SO_2 and particulate matter (per unit of mass) can take the form of equations:

$$p_{CO_2}(t) = p_{CO_2}^{t=0} e^{a_{CO_2} t} \qquad (2.30)$$

$$p_{CO}(t) = p_{CO}^{t=0} e^{a_{CO} t} \qquad (2.31)$$

$$p_{NO_x}(t) = p_{NO_x}^{t=0} e^{a_{NO_x} t} \qquad (2.32)$$

$$p_{SO_2}(t) = p_{SO_2}^{t=0} e^{a_{SO_2} t} \qquad (2.33)$$

$$p_{dust}(t) = p_{dust}^{t=0} e^{a_{dust} t} \qquad (2.34)$$

- function accounting for the cost of purchasing additional CO_2 emission allowances:

$$K_{CO_2}(t) = E_{ch,A}(1 - u) \rho_{CO_2} e_{CO_2}(t) \qquad (2.35)$$

while the fluctuation in time of the specific price of additional CO_2 emission allowances e_{CO_2} can be written in the form of an equation:

$$e_{CO_2}(t) = e_{CO_2}^{t=0} e^{b_{CO_2} t} \qquad (2.36)$$

where:

$a_{el}, a_{coal}, a_{CO_2}, a_{CO}, a_{SO_2}, a_{NO_x}, a_{dust}, b_{CO_2}$	control variables,
$E_{el,A}$	annual net production of electricity,
$E_{ch,A}$	annual use of the chemical energy of fuel,
u	ratio of chemical energy of fuel to its total use, for which it is not necessary to purchase CO_2 emission allowances,
$\rho_{CO_2}, \rho_{CO}, \rho_{NO_x}, \rho_{SO_2}, \rho_{dust}$	CO_2, CO, NO_x, SO_2 emission per unit of chemical energy of the fuel.

• function accounting for cost of maintenance and overhaul

$$K_{serv} = \delta_{serv} J \qquad (2.37)$$

where:

δ_{serv} annual rate of fixed cost dependent on the value of the investment (cost of maintenance, equipment overhaul; in practice we assume that $\delta_{serv} = 3\ \%$).

In Eqs. (2.25), (2.28), (2.30)–(2.34) and (2.36) the fluctuation of specific charges for the emission into the environment is dependent on the values of $a_{el}, a_{coal}, a_{CO_2}, a_{CO}, a_{SO_2}, a_{NO_x}, a_{dust}, b_{CO_2}$, which are strictly monotonous functions and some are constant in time.

The annual production $E_{el,A}$ in the formulae above needs to be expressed in relation to the electrical capacity of the power station N_{el}, its internal load ε_{el} and annual operating time t_A

$$E_{el,A} = N_{el}(1 - \varepsilon_{el})t_A \qquad (2.38)$$

and the annual use of the chemical energy of the fuel with the equation

$$E_{ch,A} = \frac{N_{el} t_A}{\eta_{el}} \qquad (2.39)$$

where η_{el} denotes the efficiency of electricity production in the power station.

After substituting relations (2.19), (2.22), (2.24)–(2.39) into (2.12), the solution to the problem of finding the maximum value of the functional (2.12) is limited to the trivial question of its differentiation with respect to specific boundaries and the subsequent analysis of the resulting NPV functions of independent variables: $a_{el}, a_{coal}, a_{CO_2}, a_{CO}, a_{SO_2}, a_{NO_x}, a_{dust}, b_{CO_2}$.

After a subsequent integration, we obtain:

$$
\begin{aligned}
NPV = & \left\{ N_{el}(1 - \varepsilon_{el})t_A \frac{e_{el}^{t=0}}{a_{el} - r}[e^{(a_{el}-r)T} - 1] - (1 + x_{sw,m,was})\frac{N_{el}t_A}{\eta_{el}}\frac{e_{coal}^{t=0}}{a_{coal} - r}[e^{(a_{coal}-r)T} - 1] \right. \\
& - \frac{N_{el}t_A}{\eta_{el}}\frac{\rho_{CO_2}p_{CO_2}^{t=0}}{a_{CO_2} - r}[e^{(a_{CO_2}-r)T} - 1] - \frac{N_{el}t_A}{\eta_{el}}\frac{\rho_{CO}p_{CO}^{t=0}}{a_{CO} - r}[e^{(a_{CO}-r)T} - 1] \\
& - \frac{N_{el}t_A}{\eta_{el}}\frac{\rho_{NO_x}p_{NO_x}^{t=0}}{a_{NO_x} - r}[e^{(a_{NO_x}-r)T} - 1] - \frac{N_{el}t_A}{\eta_{el}}\frac{\rho_{SO_2}p_{SO_2}^{t=0}}{a_{SO_2} - r}[e^{(a_{SO_2}-r)T} - 1] \\
& - \frac{N_{el}t_A}{\eta_{el}}\frac{\rho_{dust}P_{dust}^{t=0}}{a_{dust} - r}[e^{(a_{dust}-r)T} - 1] - \frac{N_{el}t_A}{\eta_{el}}(1 - u)\frac{\rho_{CO_2}e_{CO_2}^{t=0}}{b_{CO_2} - r}[e^{(b_{CO_2}-r)T} - 1] \\
& \left. - (1 + x_{sal,t,ins})J(1 - e^{-rT})\frac{\delta_{serv}}{r} - J_0[(1 - e^{-rT})\frac{1}{T} + 1] \right\}(1 - p).
\end{aligned}
$$

(2.40)

$x_{sw,m,}$ was	coefficient accounting for the cost of supplementing water, auxiliary materials, waste disposal, slag storage and waste (in practice, the value of $x_{sw,m,was}$ is equal to around 0.25),
$x_{sal,t,ins}$	coefficient accounting for the cost of remuneration, taxes, insurance, etc. (in practice, the value of $x_{sal,t,ins}$ is equal to around 0.02).

In the function of *NPV*, the revenues and particular costs discounted for the instance $t = 0$ are obviously rising functions within their entire range $\pm\infty$ of the variability of independent variables $a_{el}, a_{coal}, a_{CO_2}, a_{CO}, a_{SO_2}, a_{NO_x}, a_{dust}, b_{CO_2}$. Hence, all partial derivatives of the *NPV* function with respect to the specific variables (as NPV is the additive function of the discounted revenues and costs) are greater than zero ($\partial NPV/\partial a_{el} > 0$, $\partial NPV/\partial a_{coal} > 0$, $\partial NPV/\partial a_{CO_2} > 0$ etc.). This means that the *NPV* function does not have an extremum, and the variability in its values is relative to the fluctuations in the values of $a_{el}, a_{coal}, a_{CO_2}, a_{CO}, a_{SO_2}, a_{NO_x}, a_{dust}, b_{CO_2}$, that is, to the time variable relations between prices of energy carriers and environmental charges.

For the purposes of searching an optimum investment strategy, the relation (2.40) could be suitably presented as the quotient of the profit gained along the period of exploitation of the power station T per unit of power:

$$
\frac{NPV}{N_{el}} = \left\{ (1 - \varepsilon_{el}) t_A \frac{e_{el}^{t=0}}{a_{el} - r} [e^{(a_{el} - r)T} - 1] - (1 + x_{sw,m,was}) \frac{t_A}{\eta_{el}} \frac{e_{coal}^{t=0}}{a_{coal} - r} [e^{(a_{coal} - r)T} - 1] \right.
$$

$$
- \frac{t_A}{\eta_{el}} \frac{\rho_{CO_2} p_{CO_2}^{t=0}}{a_{CO_2} - r} [e^{(a_{CO_2} - r)T} - 1] - \frac{t_A}{\eta_{el}} \frac{\rho_{CO} p_{CO}^{t=0}}{a_{CO} - r} [e^{(a_{CO} - r)T} - 1]
$$

$$
- \frac{t_A}{\eta_{el}} \frac{\rho_{NO_X} p_{NO_X}^{t=0}}{a_{NO_X} - r} [e^{(a_{NO_X} - r)T} - 1] - \frac{t_A}{\eta_{el}} \frac{\rho_{SO_2} p_{SO_2}^{t=0}}{a_{SO_2} - r} [e^{(a_{SO_2} - r)T} - 1]
$$

$$
- \frac{t_A}{\eta_{el}} \frac{\rho_{dust} p_{dust}^{t=0}}{a_{dust} - r} [e^{(a_{dust} - r)T} - 1] - \frac{t_A}{\eta_{el}} (1 - u) \frac{\rho_{CO_2} e_{CO_2}^{t=0}}{b_{CO_2} - r} [e^{(b_{CO_2} - r)T} - 1]
$$

$$
\left. - (1 + x_{sal,t,ins}) i (1 - e^{-rT}) \frac{\delta_{serv}}{r} - iz[(1 - e^{-rT}) \frac{1}{T} + 1] \right\} (1 - p).
$$

$$
(2.41)
$$

where:

i specific investment (per unit of power); $i = J/N_{el}$.

Such a notation makes it much easier to perform a search since it does require input in the form of the electrical capacity of the power station N_{el}. It is only enough to describe the specific power engineering technologies—beside such specific measures relating to the power station such as annual operating time t_A, internal electrical load ε_{el}, environmental charges and cost of fuel—only in terms of the investment i necessary to pay for them.

The optimum energy strategy is the one for which the calculated value of NPV/N_{el} is the highest. This value is relative to the internal electrical load of the power station ε_{el}, its annual operating time t_A, specific investment i as well as the variability in the price relations between the energy carriers and environmental charges in time. Apparently, we are not able to predict what the variations in these price will be in the future. However, the analysis of the variability in the value NPV/N_{el} in time for various ranges of combinations in the values $a_{el}, a_{coal}, a_{CO_2}, a_{CO}, a_{SO_2}, a_{NO_X}, a_{dust}, b_{CO_2}$ enable the analysis of the future and can help us create the idea about it in a scientific manner. It will enable a rational choice to be made among the available technologies with regard to the one, which is characterized by the greatest profit, i.e. a technology which should be followed and invested in. In addition, it will guide the choice of diversifying the applied technologies as it will provide with the choice of the most effective one from the economic perspective. The diversification of the applied technologies, additionally, will increase the security of electricity supplies and will prevent the threats to the energy security of the country.

The equivalent criterion for $NPV \rightarrow$ max in the search for an optimum investment strategy in power engineering is the criterion of seeking a minimum value of the specific cost of electricity production:

$$k_{el} \rightarrow \min. \tag{2.42}$$

This cost is derived on the basis of relations (2.40) on condition that $NPV = 0$. If we additionally assume that $a_{el} = 0$, we obtain the formula expressing the mean specific cost $k_{el,av}$ [cf. formula (1.19)]:

$$
\begin{aligned}
k_{el,av} = \Bigg\{ &(1+x_{sw,m,was})\frac{e_{coal}^{t=0}}{a_{coal}-r}[e^{(a_{coal}-r)T}-1] + \frac{\rho_{CO_2}p_{CO_2}^{t=0}}{a_{CO_2}-r}[e^{(a_{CO_2}-r)T}-1] + \frac{\rho_{CO}p_{CO}^{t=0}}{a_{CO}-r}[e^{(a_{CO}-r)T}-1] \\
&+ \frac{\rho_{NO_X}p_{NO_X}^{t=0}}{a_{NO_X}-r}[e^{(a_{NO_X}-r)T}-1] + \frac{\rho_{SO_2}p_{SO_2}^{t=0}}{a_{SO_2}-r}[e^{(a_{SO_2}-r)T}-1] \\
&+ \frac{\rho_{dust}p_{dust}^{t=0}}{a_{dust}-r}[e^{(a_{dust}-r)T}-1] + (1-u)\frac{\rho_{CO_2}e_{CO_2}^{t=0}}{b_{CO_2}-r}[e^{(b_{CO_2}-r)T}-1] \\
&+ \frac{\eta_{el}}{rt_A}\left[(1+x_{sal,t,ins})i(1-e^{-rT})\delta_{serv}+riz(\frac{1-e^{-rT}}{T}+1)\right]\Bigg\} \frac{r}{\eta_{el}(1-\varepsilon_{el})(1-e^{-rT})}.
\end{aligned}
\tag{2.43}
$$

The technology which offers most economic advantages is the one for which the mean specific cost of electricity production $k_{el,av}$ is the lowest. It is, in fact, relative to the specific investment i, internal electrical load of the power station ε_{el}, its annual operating time t_A, as well as price relations between energy carriers and their variability in time, ratio of the chemical energy of fuel u for which it is not necessary to purchase CO_2 emission allowances, and on the tariffs for the use of the natural environment, etc.

By integration of the formulae in (2.13) and (2.14), we can obtain the relations for IRR rate and $DPBP$.

- Internal Rate of Return IRR

$$
\begin{aligned}
&N_{el}(1-\varepsilon_{el})t_A e_{el}^{t=0}\frac{1}{a_{el}-IRR}[e^{(a_{el}-IRR)T}-1] \\
&-\frac{N_{el}t_A}{\eta_{el}}(1+x_{sw,m,was})e_{coal}^{t=0}\frac{1}{a_{coal}-IRR}[e^{(a_{coal}-IRR)T}-1] - \frac{N_{el}t_A}{\eta_{el}}\rho_{CO_2}p_{CO_2}^{t=0}\frac{1}{a_{CO_2}-IRR}[e^{(a_{CO_2}-IRR)T}-1] \\
&-\frac{N_{el}t_A}{\eta_{el}}\rho_{CO}p_{CO}^{t=0}\frac{1}{a_{CO}-IRR}[e^{(a_{CO}-IRR)T}-1] - \frac{N_{el}t_A}{\eta_{el}}\rho_{NO_X}p_{NO_X}^{t=0}\frac{1}{a_{NO_X}-IRR}[e^{(a_{NO_X}-IRR)T}-1] \\
&-\frac{N_{el}t_A}{\eta_{el}}\rho_{SO_2}p_{SO_2}^{t=0}\frac{1}{a_{SO_2}-IRR}[e^{(a_{SO_2}-IRR)T}-1] - \frac{N_{el}t_A}{\eta_{el}}\rho_{dust}p_{dust}^{t=0}\frac{1}{a_{dust}-IRR}[e^{(a_{dust}-IRR)T}-1] \\
&-\frac{N_{el}t_A}{\eta_{el}}(1-u)\rho_{CO_2}e_{CO_2}^{t=0}\frac{1}{b_{CO_2}-IRR}[e^{(b_{CO_2}-IRR)T}-1] - J(1+x_{sal,t,ins})\frac{\delta_{serv}}{IRR}(1-e^{-IRRT}) \\
&= J\frac{(1+IRR)^{b+1}-1}{(b+1)IRR}\left(1+\frac{1-e^{-IRRT}}{T}\right)
\end{aligned}
\tag{2.44}
$$

The calculation of IRR requires the application of the method of subsequent approximations.

- Dynamic Pay-Back Period $DPBP$:

$$J_0\left[1+\frac{1}{T}-e^{-rDPBP}\left(1+\frac{1}{T}-\frac{DPBP}{T}\right)\right]$$

$$+\left\{N_{el}(1-\varepsilon_{el})t_A e_{el}^{t=0}\frac{1}{a_{el}-r}[e^{(a_{el}-r)DPBP}-1]-\frac{N_{el}t_A}{\eta_{el}}(1+x_{sw,m,was})e_{coal}^{t=0}\frac{1}{a_{coal}-r}[e^{(a_{coal}-r)DPBP}-1]\right.$$

$$-\frac{N_{el}t_A}{\eta_{el}}\rho_{CO_2}P_{CO_2}^{t=0}\frac{1}{a_{CO_2}-r}[e^{(a_{CO_2}-r)DPBP}-1]-\frac{N_{el}t_A}{\eta_{el}}\rho_{CO}P_{CO}^{t=0}\frac{1}{a_{CO}-r}[e^{(a_{CO}-r)DPBP}-1]$$

$$-\frac{N_{el}t_A}{\eta_{el}}\rho_{NO_x}P_{NO_x}^{t=0}\frac{1}{a_{NO_x}-r}[e^{(a_{NO_x}-r)DPBP}-1]-\frac{N_{el}t_A}{\eta_{el}}\rho_{SO_2}P_{SO_2}^{t=0}\frac{1}{a_{SO_2}-r}[e^{(a_{SO_2}-r)DPBP}-1]$$

$$-\frac{N_{el}t_A}{\eta_{el}}\rho_{dust}P_{dust}^{t=0}\frac{1}{a_{dust}-r}[e^{(a_{dust}-r)DPBP}-1]-\frac{N_{el}t_A}{\eta_{el}}(1-u)\rho_{CO_2}e_{CO_2}^{t=0}\frac{1}{b_{CO_2}-r}[e^{(b_{CO_2}-r)DPBP}-1]$$

$$\left.-J(1-e^{-rDPBP})(1+x_{sal,t,ins})\frac{\delta_{serv}}{r}-J_0\left[1+\frac{1}{T}-e^{-rDPBP}\left(1+\frac{1}{T}-\frac{DPBP}{T}\right)\right]\right\}(1-p)=J_0(1+\frac{1-e^{-rT}}{T})$$

$$(2.45)$$

The calculation of value of $DPBP$ requires the application of the method of subsequent approximations.

2.2.2   Mathematical Model with Continuous Time of Searching for an Optimum Investment Strategy in Sources of Combined Heat and Electricity Production

Just as in the case of sources used to produce electricity, the choice of an optimum investment strategy associated with combined heat and power sources should be made with the aim of

$$NPV \rightarrow \max \qquad (2.46)$$

or under condition that the specific cost of heat production should be the smallest:

$$k_c \rightarrow \min. \qquad (2.47)$$

This cost can be derived from the formula (2.46) when we account for the fact that beside the revenue from the sales of electricity $S_{A,el}=E_{el,A}e_{el}^{t=0}e^{a_{el}t}$ we gain an additional revenue from the sales of heat:

$$S_{A,Q}=Q_A e_c^{t=0}e^{a_c t} \qquad (2.48)$$

where:
Q_A annual net production of heat, GJ/a.

For the case of sources applying combined heat and power, this formula takes the form:

$$
\begin{aligned}
NPV = \Bigg\{ & E_{el,A}e_{el}^{t=0}\frac{1}{a_{el}-r}[e^{(a_{el}-r)T}-1] + Q_A e_c^{t=0}\frac{1}{a_c-r}[e^{(a_c-r)T}-1] \\
& -\frac{E_{el,A}+Q_A}{\eta_c}(1+x_{sw,m,was})e_{coal}^{t=0}\frac{1}{a_{coal}-r}[e^{(a_{coal}-r)T}-1] \\
& -\frac{E_{el,A}+Q_A}{\eta_c}\rho_{CO_2}p_{CO_2}^{t=0}\frac{1}{a_{CO_2}-r}[e^{(a_{CO_2}-r)T}-1] \\
& -\frac{E_{el,A}+Q_A}{\eta_c}\rho_{CO}p_{CO}^{t=0}\frac{1}{a_{CO}-r}[e^{(a_{CO}-r)T}-1] \\
& -\frac{E_{el,A}+Q_A}{\eta_c}\rho_{NO_X}p_{NO_X}^{t=0}\frac{1}{a_{NO_X}-r}[e^{(a_{NO_X}-r)T}-1] \\
& -\frac{E_{el,A}+Q_A}{\eta_c}\rho_{SO_2}p_{SO_2}^{t=0}\frac{1}{a_{SO_2}-r}[e^{(a_{SO_2}-r)T}-1] \\
& -\frac{E_{el,A}+Q_A}{\eta_c}\rho_{dust}p_{dust}^{t=0}\frac{1}{a_{dust}-r}[e^{(a_{dust}-r)T}-1] \\
& -\frac{E_{el,A}+Q_A}{\eta_c}(1-u)\rho_{CO_2}e_{CO_2}^{t=0}\frac{1}{b_{CO_2}-r}[e^{(b_{CO_2}-r)T}-1] \\
& -J(1-e^{-rT})(1+x_{sal,t,ins})\frac{\delta_{serv}}{r}-J_0[(1-e^{-rT})\frac{1}{T}+1]\Bigg\}(1-p)
\end{aligned}
$$

$$(2.49)$$

where:

η_c energy efficiency of heat and electricity production (its value is relative to the applied technology of combined heat and electricity production).

After integration of Eqs. (2.13) and (2.14), we obtain the relations expressing *IRR* and *DPBP*:

- Internal Rate of Return (*IRR*)

$$
\begin{aligned}
& Q_A e_c^{t=0}\frac{1}{a_c-IRR}[e^{(a_c-IRR)T}-1] + E_{el,A}e_{el}^{t=0}\frac{1}{a_{el}-IRR}[e^{(a_{el}-IRR)T}-1] \\
& -\frac{E_{el,A}+Q_A}{\eta_c}(1+x_{sw,m,was})e_{coal}^{t=0}\frac{1}{a_{coal}-IRR}[e^{(a_{coal}-IRR)T}-1] \\
& -\frac{E_{el,A}+Q_A}{\eta_c}\rho_{CO_2}p_{CO_2}^{t=0}\frac{1}{a_{CO_2}-IRR}[e^{(a_{CO_2}-IRR)T}-1] \\
& -\frac{E_{el,A}+Q_A}{\eta_c}\rho_{CO}p_{CO}^{t=0}\frac{1}{a_{CO}-IRR}[e^{(a_{CO}-IRR)T}-1] \\
& -\frac{E_{el,A}+Q_A}{\eta_c}\rho_{NO_X}p_{NO_X}^{t=0}\frac{1}{a_{NO_X}-IRR}[e^{(a_{NO_X}-IRR)T}-1]
\end{aligned}
$$

$$-\frac{E_{el,A}+Q_A}{\eta_c}\rho_{SO_2}p_{SO_2}^{t=0}\frac{1}{a_{SO_2}-IRR}[e^{(a_{SO_2}-IRR)T}-1]$$

$$-\frac{E_{el,A}+Q_A}{\eta_c}\rho_{dust}P_{dust}^{t=0}\frac{1}{a_{dust}-IRR}[e^{(a_{dust}-IRR)T}-1]$$

$$-\frac{E_{el,A}+Q_A}{\eta_c}(1-u)\rho_{CO_2}e_{CO_2}^{t=0}\frac{1}{b_{CO_2}-IRR_p^{IPP}}[e^{(b_{CO_2}-IRR_p^{IPP})T}-1]$$

$$-J(1+x_{sal,t,ins})\frac{\delta_{serv}}{IRR}(1-e^{-IRRT})=J\frac{(1+IRR)^{b+1}-1}{(b+1)IRR}\left(1+\frac{1-e^{-IRRT}}{T}\right)$$

$$(2.50)$$

- Dynamic Pay-Back Period (*DPBP*)

$$\left\{Q_A e_c^{t=0}\frac{1}{a_c-r}[e^{(a_c-r)T}-1]+E_{el,A}e_{el}^{t=0}\frac{1}{a_{el}-r}[e^{(a_{el}-r)DPBP}-1]\right.$$

$$-\frac{E_{el,A}+Q_A}{\eta_c}(1+x_{sw,m,was})e_{coal}^{t=0}\frac{1}{a_{coal}-r}[e^{(a_{coal}-r)DPBP}-1]$$

$$-\frac{E_{el,A}+Q_A}{\eta_c}\rho_{CO_2}p_{CO_2}^{t=0}\frac{1}{a_{CO_2}-r}[e^{(a_{CO_2}-r)DPBP}-1]$$

$$-\frac{E_{el,A}+Q_A}{\eta_c}\rho_{CO}p_{CO}^{t=0}\frac{1}{a_{CO}-r}[e^{(a_{CO}-r)DPBP}-1]$$

$$-\frac{E_{el,A}+Q_A}{\eta_c}\rho_{NO_x}p_{NO_x}^{t=0}\frac{1}{a_{NO_x}-r}[e^{(a_{NO_x}-r)DPBP}-1]$$

$$-\frac{E_{el,A}+Q_A}{\eta_c}\rho_{SO_2}p_{SO_2}^{t=0}\frac{1}{a_{SO_2}-r}[e^{(a_{SO_2}-r)DPBP}-1]$$

$$(2.51)$$

$$-\frac{E_{el,A}+Q_A}{\eta_c}\rho_{dust}P_{dust}^{t=0}\frac{1}{a_{dust}-r}[e^{(a_{dust}-r)DPBP}-1]$$

$$-\frac{E_{el,A}+Q_A}{\eta_c}(1-u)\rho_{CO_2}e_{CO_2}^{t=0}\frac{1}{b_{CO_2}-r}[e^{(b_{CO_2}-r)DPBP}-1]$$

$$\left.-J(1-e^{-rDPBP})(1+x_{sal,t,ins})\frac{\delta_{serv}}{r}\right\}(1-p)$$

$$-J_0\left[1+\frac{1}{T}-e^{-rDPBP}\left(1+\frac{1}{T}-\frac{DPBP}{T}\right)\right]p=J_0(1+\frac{1-e^{-rT}}{T})$$

The calculation of the values of *IRR* and *DPBP* from Eqs. (2.50) and (2.51) requires the application of the method of subsequent approximations.

A criterion which is equivalent to the one formulated earlier, i.e. *NPV* → max in the search for an optimum investment strategy is the one in which we seek the minimum value of the specific cost of heat production:

$$k_c \rightarrow \min.$$

$$(2.52)$$

This cost is derived from the relation in (2.49) on condition that $NPV = 0$:

$$Q_A k_c^{t=0} \frac{1}{a_c - r} [e^{(a_c - r)T} - 1] = \frac{E_{el,A} + Q_A}{\eta_c} (1 + x_{sw,m,was}) e_{coal}^{t=0} \frac{1}{a_{coal} - r} [e^{(a_{coal} - r)T} - 1]$$

$$+ \frac{E_{el,A} + Q_A}{\eta_c} \rho_{CO_2} p_{CO_2}^{t=0} \frac{1}{a_{CO_2} - r} [e^{(a_{CO_2} - r)T} - 1]$$

$$+ \frac{E_{el,A} + Q_A}{\eta_c} \rho_{CO} p_{CO}^{t=0} \frac{1}{a_{CO} - r} [e^{(a_{CO} - r)T} - 1]$$

$$+ \frac{E_{el,A} + Q_A}{\eta_c} \rho_{NO_X} p_{NO_X}^{t=0} \frac{1}{a_{NO_X} - r} [e^{(a_{NO_X} - r)T} - 1]$$

$$+ \frac{E_{el,A} + Q_A}{\eta_c} \rho_{SO_2} p_{SO_2}^{t=0} \frac{1}{a_{SO_2} - r} [e^{(a_{SO_2} - r)T} - 1]$$

$$+ \frac{E_{el,A} + Q_A}{\eta_c} \rho_{dust} p_{dust}^{t=0} \frac{1}{a_{dust} - r} [e^{(a_{dust} - r)T} - 1]$$

$$+ \frac{E_{el,A} + Q_A}{\eta_c} (1 - u) \rho_{CO_2} e_{CO_2}^{t=0} \frac{1}{b_{CO_2} - r} [e^{(b_{CO_2} - r)T} - 1]$$

$$+ J(1 - e^{-rT})(1 + x_{sal,t,ins}) \frac{\delta_{serv}}{r} + J_0 [(1 - e^{-rT}) \frac{1}{T} + 1]$$

$$- E_{el,A} e_{el}^{t=0} \frac{1}{a_{el} - r} [e^{(a_{el} - r)T} - 1]$$

$$(2.53)$$

and for $a_c = 0$, we can derive the mean specific cost of heat production [cf. formula (1.18)]:

$$k_{c,av} = \frac{\sigma_A + 1}{\eta_c (1 - e^{-rT})} \left\{ (1 + x_{sw,m,was}) e_{coal}^{t=0} \frac{r}{a_{coal} - r} [e^{(a_{coal} - r)T} - 1] \right.$$

$$+ \rho_{CO_2} p_{CO_2}^{t=0} \frac{r}{a_{CO_2} - r} [e^{(a_{CO_2} - r)T} - 1]$$

$$+ \rho_{CO} p_{CO}^{t=0} \frac{r}{a_{CO} - r} [e^{(a_{CO} - r)T} - 1]$$

$$+ \rho_{NO_X} p_{NO_X}^{t=0} \frac{r}{a_{NO_X} - r} [e^{(a_{NO_X} - r)T} - 1]$$

$$+ \rho_{SO_2} p_{SO_2}^{t=0} \frac{r}{a_{SO_2} - r} [e^{(a_{SO_2} - r)T} - 1] \qquad (2.54)$$

$$+ \rho_{dust} p_{dust}^{t=0} \frac{r}{a_{dust} - r} [e^{(a_{dust} - r)T} - 1]$$

$$\left. + (1 - u) \rho_{CO_2} e_{CO_2}^{t=0} \frac{r}{b_{CO_2} - r} [e^{(b_{CO_2} - r)T} - 1] \right\}$$

$$+ i(1 + x_{sal,t,ins}) \delta_{serv}$$

$$+ \frac{rzi}{(1 - e^{-rT})} [(1 - e^{-rT}) \frac{1}{T} + 1] - \frac{r \sigma_A e_{el}^{t=0}}{(a_{el} - r)(1 - e^{-rT})} [e^{(a_{el} - r)T} - 1]$$

where:

$\sigma_A = \frac{E_{el,A}}{Q_A} \geq 0$ annual cogeneration index (its value is relative to the application of a specific technology of combined heat and electricity production, the highest value is nowadays noted for heat and power plant in the gas-steam technology; $\sigma_A^{G-S} \cong 4,1$ [3]).

Equation (2.54) can be written in dimensionless form by dividing its both sides e.g. by the specific price of electricity $e_{el}^{t=0}$. As a result, it represents the average specific cost of heat production $k_{c,av}$ related to the specific cost of electricity $e_{el}^{t=0}$ only in the function of dimensionless independent variables: $\sigma_A, a_{el}, a_{coal}, a_{CO_2}, a_{CO}, a_{SO_2},$ $a_{NO_x}, a_{dust}, b_{CO_2}, u, \eta_c, e_{coal}^{t=0}/e_{el}^{t=0}$ etc., $\rho_{CO_2} p_{CO_2}^{t=0}/e_{el}^{t=0}$ etc., $i/e_{el}^{t=0} = J/(e_{el}^{t=0}Q_A)$:

$$
\begin{aligned}
\frac{k_{c,av}}{e_{el}^{t=0}} = & \frac{\sigma_A+1}{\eta_c(1-e^{-rT})} \left\{ (1+x_{sw,m,was}) \frac{e_{coal}^{t=0}}{e_{el}^{t=0}} \frac{r}{a_{coal}-r} [e^{(a_{coal}-r)T}-1] \right. \\
& + \rho_{CO_2} \frac{p_{CO_2}^{t=0}}{e_{el}^{t=0}} \frac{r}{a_{CO_2}-r} [e^{(a_{CO_2}-r)T}-1] \\
& + \rho_{CO} \frac{p_{CO}^{t=0}}{e_{el}^{t=0}} \frac{r}{a_{CO}-r} [e^{(a_{CO}-r)T}-1] + \rho_{NO_x} \frac{p_{NO_x}^{t=0}}{e_{el}^{t=0}} \frac{r}{a_{NO_x}-r} [e^{(a_{NO_x}-r)T}-1] \\
& + \rho_{SO_2} \frac{p_{SO_2}^{t=0}}{e_{el}^{t=0}} \frac{r}{a_{SO_2}-r} [e^{(a_{SO_2}-r)T}-1] + \rho_{dust} \frac{p_{dust}^{t=0}}{e_{el}^{t=0}} \frac{r}{a_{dust}-r} [e^{(a_{dust}-r)T}-1] \\
& \left. + (1-u)\rho_{CO_2} \frac{e_{CO_2}^{t=0}}{e_{el}^{t=0}} \frac{r}{b_{CO_2}-r} [e^{(b_{CO_2}-r)T}-1] \right\} + (1+x_{sal,t,ins}) \frac{i\delta_{serv}}{e_{el}^{t=0}} \\
& + \frac{rzi}{e_{el}^{t=0}(1-e^{-rT})} [(1-e^{-rT})\frac{1}{T}+1] - \frac{r\sigma_A}{(a_{el}-r)(1-e^{-rT})} [e^{(a_{el}-r)T}-1]
\end{aligned}
$$

$$(2.55)$$

The dimensionless form of Eq. (2.55) is very suitable for analyzing the variability in the cost of heat and electricity production. It has taken on a very general form and enabled the analysis of technical and economic efficiency to be performed with regard to combined heat and power sources regardless of their thermal capacity and technology. In other words, it provides grounds for the transfer of the results of calculations into sources with various technical and economic parameters.

The value of the dimensionless ratio $k_{c,av}/e_{el}^{t=0}$ can assume a negative value since the cost $k_{c,av}$ can be negative due to the avoided cost which is equal with the minus sign to the revenue from the sales of electricity sold from power source: $-E_{el,A}e_{el}^{t=0}\frac{1}{a_{el}-r}[e^{(a_{el}-r)T}-1]$ (formula 2.53). The maximum value of this ratio should be considerably smaller than one. If the specific cost of heat was close to the cost of electricity, the system with the heat and power plant would be completely unjustified both for both economic as well as thermodynamic purposes.

An optimum technology is the one, in which the specific relative cost of heat production is the smallest. It is, in turn, relative to the annual electricity output $E_{el,A}$

in relation to the annual production of heat Q_A, that is, on the annual value of the σ_A parameter (which, as noted before, it relative to the applied technology) and on the price relations between energy carriers and their variability in time, i.e. on the relation of fuel prices (coal, gas) to the price of electricity and the charges associated with the use of the natural environment.

After performing calculations involving several alternatives in accordance with formula (2.55), it will be possible to present them in a graphical form, for instance by means of nomograms with the specific cost of heat production $k_{c,av}/e_{el}^{t=0}$ by means of dimensionless measures σ_A, $e_{coal}^{t=0}/e_{el}^{t=0}$ etc., $i/e_{el}^{t=0} = J/(e_{el}^{t=0}Q_A)$ etc. as the parameters. Due to the use of such nomograms, it will be possible each time to identify the optimum technology of combined heat and electricity production for the specific prices of energy carrier prices and charges associated with the use of environment as well as their forecasted variability in time. Subsequently, it will be possible to relate these variables to the values of thermal power of a heat and power plant.

In summary, the presented methodology and model used to perform technical and economic analysis of a search for an optimum investment strategy in heat sources offers both cognitive values and offers the option to broaden our knowledge regarding investment strategies. This option, in turn, provides a wide range of applications and activities.

2.3 Results of Sample Calculations

The chapter reports the results of a comparative analysis of the specific cost of electricity production in a conventional power station with supercritical parameters with a power station utilizing oxy-fuel combustion.

The utilization of oxy-fuel technology in power units is an idea guided by the necessity imposed by the need to fulfill EU regulations regarding the reduction of CO_2 emission into the atmosphere. One of the CCS (Carbon Capture and Storage) technologies in this area involves the process of coal combustion in a power plant using pure oxygen instead of air as the primary oxidant, i.e. in the oxy-fuel process—Fig. 2.5.

After the combustion process, the gases at the exhaust stage of the boiler in this case are cooled and H_2O is condensed. The gas from this process contains nearly pure carbon dioxide, which can be transported to its destination for storage in deep rock formations. The analysis presents in the chapter involves a power unit with supercritical parameters with the electrical capacity of $N_{el} = 460$ MW and gross efficiency of $\eta_{el} = 45.6\,\%$. For the operation of the unit in the oxy-fuel combustion it was additionally supplemented by a cryogenic oxygen station—Fig. 2.5. In the comparative calculations formula (2.43) has been used.

For the case of the oxy-fuel process, the cost K_{env} and K_{CO_2} (formulas 2.29 and 2.35) does not occur in the total exploitation cost K_e (formula 2.26). Instead of them, it is necessary to include the cost K_{CCS} associated with the transport of carbon

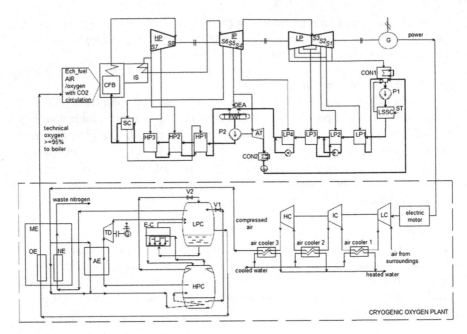

Fig. 2.5 Diagram of power unit with supercritical parameters utilizing oxy-fuel combustion repowered by two-column cryogenic oxygen station powered by electric motor: *AE* auxiliary heat exchanger, *AT* auxiliary turbine, *CFB* circulating fluidized bed boiler, *CON1* main condenser, *CON2* auxiliary turbine condenser, *DEA* deaerator, *E-C* heat exchanger "evaporator–condenser", *Ech_fuel* chemical energy of fuel, *FWT* feed water tank, *G* electricity generator, *HC* high pressure compressor section, *HP* high pressure section of the steam turbine, *HP1-HP3* high pressure preheaters, *HPC* high pressure rectification column, *IC* intermediate pressure compressor section, *IP* intermediate pressure section of the steam turbine, *IS* interstage superheater, *LC* low pressure compressor section, *LP* low pressure section of the steam turbine, *LP1-4* low pressure preheaters, *LPC* low pressure rectification column, *LSSC* labyrinth seal steam condenser, *ME* main heat exchanger for cooling the compressed air using technical oxygen (OE) and waste nitrogen (NE), *P1* condensate pump, *P2* feed water pump, *S1-S8* bleeding steam from the main turbine, *SC* steam cooler, *ST* steam from labyrinth sealings in the turbine, *TD* turbodetander (turbine expander), *V1, V2* throttling valves

dioxide to the destination of its storage (formula 2.56). However, the exact input value associated with the cost K_{CCS} is possible only in the case when the place of its storage and the cost of transport are already known. The cost K_{CCS} is relative to the fact whether the transport of CO_2 occurs in a liquid form, i.e. at a temperature which is considerably below the ambient temperature in isolated tanks on tankers, by rail or by road freight, or if it is transported over a pipeline in a gaseous state. In this case it is necessary to take into account the length and diameter of the pipeline, as well as the pressure to which the CO_2 needs to be compressed above the critical pressures so that a phase change does not occur and its sufficient density is guaranteed. The way of the transport over a pipeline is more effective from the perspective energy efficiency and, hence, it can be presumed to be more economical.

The inability to determine the cost K_{CCS} without the input in the form of the above data leads to the assumption that the cost could be assumed to be proportional to the cost of K_{coal} for the comparative analysis, as it forms the most basic component of the exploitation cost K_e. Hence, we can obtain the fraction of the total cost x_{CCS} which can be attributed to this variable

$$K_{CCS} = x_{CCS}K_{coal} \qquad (2.56)$$

By dividing the cost K_{CCS} by the annual net production of electricity $E_{el,A}$ and the annual mass of carbon dioxide generated in the power unit $E_{ch,A}\rho_{CO_2}$ we can derive the formulae for the specific cost of its transport into the place of its pumping and storage expressed, respectively, per specific unit of electricity production and unit of carbon dioxide mass (Fig. 2.8):

$$k_{el,CCS} = \frac{K_{CCS}}{E_{el,A}} \qquad (2.57)$$

$$k_{CO_2,CCS} = \frac{K_{CCS}}{E_{ch,A}\rho_{CO_2}} \qquad (2.58)$$

The specific cost $k_{el,CCS}$ apparently forms a component of its specific cost $k_{el,oxy}$. The calculations should adopt a variable value of proportionality coefficient x_{CCS} in a relatively wide range, e.g. $x_{CCS} \in \langle 0; 1 \rangle$. Such an approach makes it possible to assess the influence of the cost K_{CCS} on the economic profitability of the design, construction and exploitation of the power unit in the oxy-fuel technology, regardless of the way of transport and distance of CO_2 storage. The *necessary condition* for this profitability is associated with the fulfillment of the relation that this specific cost of electricity production in a power unit realizing oxy-fuel process will not be higher than the cost of production in a conventional power unit, i.e. one with the combustion of fuel in air atmosphere:

$$k_{el,oxy} \leq k_{el} \qquad (2.59)$$

The sufficient condition is associated with the fact that the annual profit from the exploitation of the power unit secures a sufficiently large increase of the discounted net present value NPV_{oxy}, relatively short pay-back period $DPBP_{oxy}$ of the investment J_{oxy} and relatively high internal rate of return IRR_{oxy}, which is higher than the rate r [2]. The fulfillment of these conditions is to a large degree relative to the price e_{CO_2} of the CO_2 emission allowances, proportion of the chemical energy of fuel u in its total annual use, for which is it not necessary to purchase CO_2 emission permits and to the cost K_{CCS}.

Figures 2.6, 2.7 and 2.8 contain curves, for which the values of the controls were taken to be equal to zero, i.e. $a_{el} = 0, a_{coal} = 0, a_{CO_2} = 0, a_{CO} = 0, a_{SO_2} = 0,$ $a_{NO_X} = 0, a_{dust} = 0, b_{CO_2} = 0$. This means that all prices [in formulae (2.25), (2.28), (2.30)–(2.34) and (2.36)] are constant over the entire period T when the

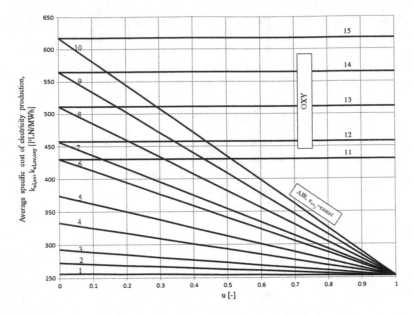

Fig. 2.6 Average specific cost of electricity production in the function of the proportion of chemical energy of coal combustion u in its total use for which it is not required to purchase CO_2 emission allowances, where: 1 $k_{el,av}$ for $e_{CO_2} = 5$ PLN/Mg_{CO_2}; 2 $k_{el,av}$ for $e_{CO_2} = 25$ PLN/Mg_{CO_2}; 3 $k_{el,av}$ for $e_{CO_2} = 50$ PLN/Mg_{CO_2}; 4 $k_{el,av}$ for $e_{CO_2} = 100$ PLN/Mg_{CO_2}; 5 $k_{el,av}$ for $e_{CO_2} = 150$ PLN/Mg_{CO_2}; 6 $k_{el,av}$ for $e_{CO_2} = 219$ PLN/Mg_{CO_2}; 7 $k_{el,av}$ for $e_{CO_2} = 252$ PLN/Mg_{CO_2}; 8 $k_{el,av}$ for $e_{CO_2} = 319 PLN/Mg_{CO_2}$; 9 $k_{el,av}$ for $e_{CO_2} = 385$ PLN/Mg_{CO_2}; 10 $k_{el,av}$ for $e_{CO_2} = 450$ PLN/Mg_{CO_2} and in the oxy-fuel technology, where: 11 $k_{el,av,oxy}$ for $x_{ccs} = 0$; 12 $k_{el,av,oxy}$ for $x_{ccs} = 0.2$; 13 $k_{el,av,oxy}$ for $x_{ccs} = 0.6$; 14 $k_{el,av,oxy}$ for $x_{ccs} = 1$; 15 $k_{el,av,oxy}$ for $x_{ccs} = 1.4$

power station operates. The calculations have also adopted the following input data included in the Table 2.1.

Figure 2.6 illustrates the average specific cost of electricity production $k_{el,av}$ and $k_{el,av,oxy}$ in the function of the proportion u of the chemical energy of the fuel in its total annual use, for which is not necessary to purchase CO_2 emission permits and with the prices e_{CO_2} of these allowances as the parameters of the equation. As the curves indicate, the fulfillment of relation (2.59) will be hard to achieve. Even if the cost $k_{el,av,CCS}$ would be avoided (Fig. 2.8), i.e. when $x_{CCS} = 0$, then the fulfillment of the relation $k_{el,av,oxy} \leq k_{el,av}$ will only be possible after the e_{CO_2} price of the purchase of CO_2 emission allowances of $e_{CO_2} \geq 219$ PLN/Mg_{CO_2} is exceeded (Fig. 2.7), which seems quite improbable to occur in the near future. The cost $k_{el,av,oxy} = k_{el,av} = 430$ PLN/MWh would in this case be higher than the cost of generating electricity in nuclear power stations, as it is equal to around 420 PLN/MWh [2]. In such circumstances, the design and building of power units with *CCS* installations would not serve any economic purpose. In reality $k_{el,av,CCS} > 0$, and hence, $x_{CCS} > 0$ and the minimum value of e_{CO_2} e.g., for the value of $x_{CCS} = 0.2$ would be even higher and equal to $e_{CO_2} = 252$ PLN/Mg_{CO_2}; as a consequence,

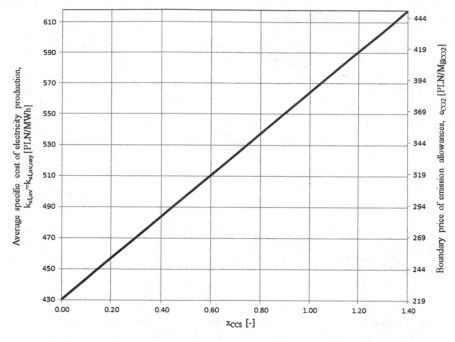

Fig. 2.7 Boundary price of e_{CO_2} emission allowances in the function of x_{CCS}, for which the average specific cost of electricity production will be brought to an equal level for air and oxy-fuel process $k_{el,av,oxy} = k_{el,av}$ for the proportion of the chemical energy of the fuel in its total use equal to u = 0

$k_{el,av,oxy} = k_{el,av} = 457$ PLN/MWh (Fig. 2.7), which would even further disqualify power units operating in the *CCS* technology. For $x_{CCS} = 1$, the boundary price of purchase of one ton of carbon dioxide emission allowance would be equal to as much as $e_{CO_2} = 385$ PLN/Mg$_{CO_2}$ and $k_{el,av,oxy} = k_{el,av} = 565$ PLN/MWh (Fig. 2.7).

Figure 2.7. presents the values of boundary values e_{CO_2} associated with carbon dioxide allowances in the function of x_{CCS}, i.e. the prices above which it is profitable to develop power units with *CCS* installations. The presentation also includes the average cost $k_{el,av,oxy}, k_{el,av}$ corresponding to these prices, which are then equal to each other, i.e. $k_{el,av,oxy} = k_{el,av}$. Prices e_{CO_2} and costs $k_{el,av,oxy} = k_{el,av}$ are presented for the value u equal to zero, $u = 0$, as in 2020 power plants will not receive free CO_2 emission allowances and it will be necessary to pay for every ton of carbon dioxide emissions.

Figures 2.9 and 2.10 illustrate the results of the calculation of average specific cost of electricity production for the variable in time prices of coal purchase and CO_2 emission allowances (formulae 2.28 and 2.36) over the entire period T when the power station operates. The initial price of coal was adopted at a level of $e_{coal}^{t=0} = 11.4$ PLN/GJ and initial price of purchasing CO_2 emission allowances equal to $e_{CO_2}^{t=0} = 29.4$ PLN/Mg$_{CO_2}$. The calculations were also performed for the ratio of

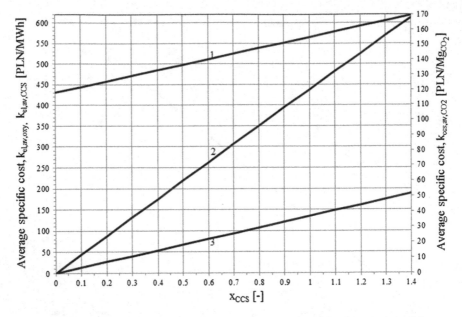

Fig. 2.8 Average specific cost of electricity production in the function of x_{CCS}, where: *1* $k_{el,av,oxy}$ [PLN/MWh]; *2* $k_{CCS,av,CO2}$ [PLN/Mg$_{CO2}$]; *3* $k_{el,av,\ CCS}$ [PLN/MWh]

Table 2.1 Input data for the calculations	
	Estimated specific investment in a power unit i = 6.5 m PLN/MW
	Estimated specific investment in a power unit including cryogenic oxygen station i_{oxy} = 9.1 mln PLN/MW
	Annual operating time of a power unit t_A = 7500 h/a
	Internal load of the power unit: ε_{el} = 7.6 %, $\varepsilon_{el,oxy}$ = 33 %
	Discount rate r = 8 %
	Period needed for construction of a power unit b = 5 years
	Exploitation period of a power unit T = 20 years
	Annual rate of maintenance and overhauls δ_{serv} = 3 %
	Total of cost $K_{sal} + K_t = 0.25 K_{serv}$, total of cost $K_{sw} + K_m = 0.02 K_{coal}$
	Specific coal price e_{coal} = 11.4 PLN/GJ
	Price of CO_2 emission allowances: e_{CO_2} = 29.4 PLN/Mg$_{CO_2}$ (7 euro; EURO/PLN rate = 4.2)
	Specific tariffs per emission of substances: p_{CO_2} = 0.29 PLN/Mg$_{CO_2}$, p_{CO} = 110 PLN/Mg$_{CO}$, p_{NO_x} = 530 PLN/Mg$_{NO_x}$, p_{SO_2} = 530 PLN/Mg$_{SO_2}$, p_{dust} = 350 PLN/Mg$_{pyø}$
	Emissions associated with coal combustion: ρ_{CO_2} = 95 kg/GJ, ρ_{CO} = 0.01 kg/GJ, ρ_{NO_x} = 0.164 kg/GJ, ρ_{SO_2} = 0.056 kg/GJ, ρ_{dust} = 0.007 kg/GJ

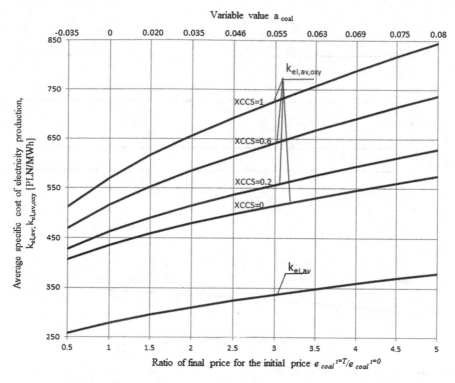

Fig. 2.9 Average specific cost of electricity production in the function of the fuel price $e^{t=T}_{coal}/e^{t=0}_{coal}$ ratio, where: $k_{el,av}$ average cost for power station for the case of air combustion technology; $k_{el,av,oxy}$ average cost for power station for the case of oxy-fuel technology

the chemical energy of coal for which it is not necessary to purchase CO_2 allowances in its total use equal to zero ($u = 0$).

From the courses in Figs. 2.6, 2.7, 2.8, 2.9 and 2.10, one can conclude that the investment in a power station with *CCS* installation is highly unprofitable from the economic perspective due to the high cost of producing electricity in it. What is more, it is not very probable that such an investment could ever make economic and technical sense. This results from the lower cost of production in a nuclear power station, even accounting for the two times higher specific investment necessary to build nuclear units ($i_{atom} \approx 18$ mln PLN/MW). Moreover, in the long term, i.e. after depreciation of the investment, the electricity from the nuclear power source will be considerably cheaper from the power production in a depreciated power unit designed for operation in supercritical parameters, even without *CCS* installation. In such a case, the price of electricity will be decided merely by the cost of the nuclear fuel, which constitutes only a small fraction of the specific cost in relation to the revenue gained from the production of electricity. The same proportion in the coal-fired power units is equal to several dozen per cent.

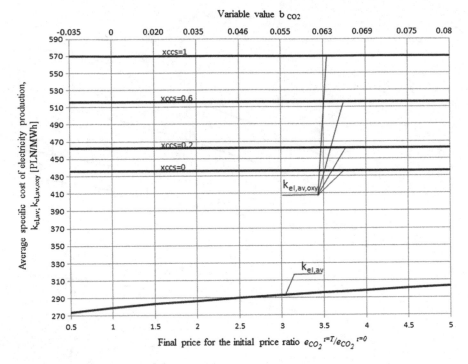

Fig. 2.10 Average specific cost of eletricity production in relation to the ratio of the prices associated with CO_2 emission $e_{CO_2}^{t=T}/e_{CO_2}^{t=0}$, where: $k_{el,av}$ average cost for power station for the case of air combustion technology; $k_{el,av,oxy}$ average cost for power station for the case of oxy-fuel technology

2.4　Conclusions

The application of a specific technology and its solutions determine the amount that is needed for an investment J_0 connected with building of a power station (in a general case this may be a source of electricity and heat). Hence, it decides on the value of the cost of finance F and loan installments R which determine the annual operating costs of a power station and along with the prices of energy carriers and specific charges for pollutant emissions into the environment decide on the annual revenue S_A and annual exploitation costs K_e. Therefore, the selection of technology decides on the value of *NPV*. An optimum strategy regarding the technology will therefore be associated with the selection of a single one for which the value of *NPV* gains a maximum value for an adopted values of the overall electrical capacity N_{el} of a power plant or in case of heat and power plant for an adopted values of combined cycle ratio.

At the same time one can note that investment decisions are long-term ones and inseparably are connected with a risk of failure. The assessment of time and the risk associated with it is difficult and sometimes impossible to forecast, in particular in unstable economic situation. This risk, however, should not discourage the investor from the search for an optimum investment strategy. The search for it can enable one to analyze the resulting description of the prospects. This can also provide means to analyze the future in a scientific manner, i.e. in a way that involves the considerations of variable price relations between energy carriers, costs associated with the use of the environment, all factors which can affect the decision regarding the selection of a strategy. The results of searching for a maximum of the functional (2.12) should therefore provide information how the price relation and environmental charges influence the selection of an investment strategy and optimum energy technology.

A means which can minimize the risk can be associated with diversification of the technology, which also involves the need to analyze the future prospects. In contrast, this can lead to a rational decision regarding diversification of the applied technologies to ensure that the most effective one from the economic perspective is selected. As a result, the security of the supply of electricity can be additionally improved.

In summary, the application of mathematical models in economics and their analysis by means of forward thinking scenarios will consequently enable the rational selection of a strategy to be adopted in order to ensure that the desired characteristics can be achieved in an optimum manner.

References

1. Korn, G.A., Korn, T.M.: Mathematical Handbook for Scientists and Engineers. McGraw-Hill Book Company, New York (1963)
2. Bartnik, R., Bartnik, B.: Economic Calculus in Power Industry (Wydawnictwo Naukowo-Techniczne WNT), Warszawa (2014)
3. Bartnik, R.: Combined Cycle Power Plants. Thermal and Economic Effectiveness (Wydawnictwo Naukowo-Techniczne WNT), Warszawa (2009) (reprint 2012)

Chapter 3
Value of the Heat and Electricity Market and Market Worth of Power Stations and Heat and Power Plants Supplying the Market

Abstract Presented is a specific methodology based on continuous time recording, of the heat and electric energy market value complex analysis as well as of the market value determination of the being sold (under privatization) power stations and CHP plants and the newly built energy sources. The essence of the market value method consists in implementing not only the future cash flows into the discount account but also the so-called relative market value. The market value methodology allows the investor to calculate the total profit that would be gained from a power plant operation, discounted return period of financial resources invested by him into the purchase of the existing ones or building of new energy sources as well as to calculate the interest rate that will be brought by the capital invested into the sources.

3.1 Introduction

The value of the market is the profit expressed financially which can be gained from a market by investing financial resources in it.

The study in [1] focuses on discreet models used for the assessment of the value of the heat and electricity market and market worth of power stations and heat and power plants. The development of these models applies discounting (dynamic) methods for the assessment of economic efficiency of the business processes, i.e. ones which account for the change of the value of money in time and accounting for benefits in terms of cash flows. Because of their precision, discounting methods are considered to be more effective criteria for the assessment of economic efficiency of phenomena and economic processes than traditional methods, i.e. ones which do not consider the change in the value of money in time. Among the discounting methods, the ones which involve the least drawbacks are the ones based on Net Present Value (NPV), Internal Rate of Return (IRR) and Dynamic Pay-Back Period (DPBP) [1]. If these models are input with functions, in which one of the arguments

R. Bartnik et al., *Optimum Investment Strategy in the Power Industry*,
SpringerBriefs in Applied Sciences and Technology,
DOI 10.1007/978-3-319-31872-1_3

is time, then such models can give an idea about the investment in a scientific manner and have the power of prediction. The results of calculation based on *NPV*, *IRR*, *DPBP* ratios and their subsequent analysis results leads to rational investment in power engineering.

The fundamental criterion for the assessment value of the heat and electricity market and market worth of energy sources should be based on the *IRR* rate gained from the invested capital. In contrast to other parameters, this measure enables the investor to assess the economic efficiency of the capital invested in the purchase and modernization of existing power stations and heat and power plants or construction of new ones including the interest which could be gained from the allocation of financial resources in the capital market. The values of *NPV* and *DPBP* are measures based on the results. The sufficiently high rate of *IRR* of the invested capital offers and incentive for *IPPs* (*Independent Power Producers*) to invest in the purchase of existing energy sources from former owners (e.g. privatization of enterprises by the State Treasury) and construction of new ones. The independent power producers mentioned above are willing to invest if they have a guarantee to sell electricity and heat in a longer time perspective. Usually an investor imposes a minimum level of IRR^{IPP} rate, which needs to be brought by the invested capital. Such a minimum, threshold value of the IRR^{IPP} rate, which compensates for be risk of investing resources is set by the local investors usually at a level of a ten or so per cent while the foreign investors expect a higher one. This rate forms an incentive to invest side by side with the long-term contracts for the sale as well as price for the goods and services produced. This rate is believed to be higher along with the level of the risk associated with the local market and democracy hazard forecasted in a specific market.

This chapter presents the original methodology using continuous notation applied for complex analysis of the value of the of heat and electricity market, and of valuation of the market worth of existing privatised heat and power plants as well as newly built energy sources. The essence of the *market worth* involves taking into account the discount from, not only future cash flows but also from so call relative value of the market v_m [1]. The method applying the notion of the *market worth* allows the *IPP* investor calculation of the total NPV^{IPP} profit obtained from operation of the existing heat and power plants purchased from the previous owners and Discounted Payback Period $DPBP^{IPP}$ of the financial resources invested into such a purchase or into construction of the new energy sources, as well as the calculation of interest rate of IRR^{IPP} expected to be achieved from the invested capital.

3.2   Methodology of Analysis and Assessment of the Value of the Heat and Electricity Market and Market Worth of Power Stations and Heat and Power Plants

By application of geometric series, monograph [1] shows the methodology and resulting discrete models of heat and electricity market as well as models of market worth of power stations and heat and power plants.

The methodology is described by the formulae below [which forms a generalization of the formulae (1.1)–(1.3)]:

- The total profit gained by an IPP investor from the operation of purchased and modernized company:

$$NPV^{IPP} = \sum_{t=1}^{M} \frac{CF_{t\,net}^{M\,IPP}}{(1+r)^t} + \frac{CF_{A\,net}^{M+1\,IPP}}{(1+r)^{M+1}} + \sum_{t=M+2}^{N} \frac{CF_{t\,net}^{mod\,IPP}}{(1+r)^t} - J_0 - \frac{J_M}{(1+r)^M}$$

(3.1)

- Internal Rate of Return IRR^{IPP} gained on the investment J_0 and J_M in the purchase and modernization:

$$\sum_{t=1}^{M} \frac{CF_{t\,gross}^{M\,IPP}}{(1+IRR^{IPP})^t} + \frac{CF_{R\,gross}^{M+1\,IPP}}{(1+IRR^{IPP})^{M+1}} + \sum_{t=M+2}^{N} \frac{CF_{t\,gross}^{mod\,IPP}}{(1+IRR^{IPP})^t}$$
$$= J_0 + \frac{J_M}{(1+IRR^{IPP})^M}$$

(3.2)

- Discounted (dynamic) Pay-Back Period $DPBP^{IPP}$ associated with the investment J_0 and J_M:

$$\sum_{t=1}^{M} \frac{CF_{t\,net}^{M\,IPP}}{(1+r)^t} + \frac{CF_{A\,net}^{M+1\,IPP}}{(1+r)^{M+1}} + \sum_{t=M+2}^{DPBP^{IPP}} \frac{CF_{t\,net}^{mod\,IPP}}{(1+r)^t} = J_0 + \frac{J_M}{(1+r)^M}$$

(3.3)

where:

J_0 purchase price of a company from the previous owner,

J_M investment expenditures in year M necessary for renovation and modernization of the purchased company for price J_0; J_M is a function of the technical condition of existing equipment as well as of the extent and technique of modernization,

while the cash flows in the subsequent years (Fig. 3.1) are:

- gross cash flow

$$\text{in years } 1, 2, \ldots, M \quad CF_{A\,gross}^{M\,IPP} = \rho_N J_0 + Z^M(1 - v_m) \quad (3.4)$$

$$\text{in year } M+1 \quad CF_{A\,gross}^{M+1\,IPP} = \rho_N J_0 + \rho_{N-M} J_M + Z^{M+1}(1 - v_m) \quad (3.5)$$

$$\text{in years } M+2, \ldots, N \quad CF_{A\,gross}^{mod\,IPP} = \rho_N J_0 + \rho_{N-M} J_M + Z^{mod}(1 - v_m) \quad (3.6)$$

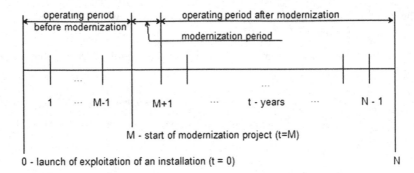

Fig. 3.1 Time line representing operation of a company

- net cash flow

$$\text{in years } 1, 2, \ldots, M \quad CF_{A\,net}^{M\,IPP} = \rho_N J_0 + Z^M (1 - p)(1 - v_m) \tag{3.7}$$

$$\text{in year } M + 1 \quad CF_{A\,net}^{M+1\,IPP} = \rho_N J_0 + \rho_{N-M} J_M + Z^{M+1}(1 - p)(1 - v_m) \tag{3.8}$$

$$\text{in years } M + 2, \ldots, N \quad CF_{A\,net}^{mod\,IPP} = \rho_N J_0 + \rho_{N-M} J_M + Z^{mod}(1 - p)(1 - v_m) \tag{3.9}$$

where:
v_m relative market value of goods and services produced by a company, expressed as a percentage of the market share held by a particular player of the market, e.g. The State Treasury [percentage value of the total profit earnt during the period of operation of company NPV, formula (1.1); cf. formula (3.12)],
ρ rate of depreciation including interest,
P tax rate on profit before tax Z [formula (1.12); cf. formula (3.14)].

The formula in (3.2) gives the gross cash flows. The net annual profit $Z(1 - p)$ $(1 - v_m)$, which is gained by an IPP investor, is the value after tax is deduced and the investor can request that the IRR^{IPP} value for this profit will be calculated for the value of this net profit. Therefore, taking into account the income tax P on gross profit Z, the internal rate of return IRR_p^{IPP} should be calculated from:

$$\sum_{t=1}^{M} \frac{CF_{t\,net}^{M\,IPP}}{(1 + IRR_p^{IPP})^t} + \frac{CF_{A\,net}^{M+1\,IPP}}{(1 + IRR_p^{IPP})^{M+1}} + \sum_{t=M+2}^{N} \frac{CF_{t\,net}^{mod\,IPP}}{(1 + IRR_p^{IPP})^t}$$
$$= J_0 + \frac{J_M}{(1 + IRR_p^{IPP})^M} \tag{3.10}$$

Equation (3.10) can be applied to find the purchase price J_0 of a power station or heat or power plant from the previous owner (e.g. from the State Treasury) as a function of IRR_p^{IPP} at the rate demanded by the IPP investor [rates include the income tax from gross profit Z, see Eq. (1.12)], with relative market value v_m and the necessary investment expenditure on overhaul and modernization J_M borne by the IPP investor as the parameters [cf. formula (3.24)]:

$$
J_0 = \frac{\left[\frac{(S_A^M - K_e^M)}{[\rho_M]_{IRR_p^{IPP}}} + \frac{(S_A^{M+1} - K_e^{M+1})}{(1 + IRR_p^{IPP})^{M+1}} + (S_A^{mod} - K_e^{mod}) \left(\frac{1}{[\rho_N]_{IRR_p^{IPP}}} - \frac{1}{[\rho_{M+1}]_{IRR_p^{IPP}}} \right) \right] (1 - p)(1 - v_m)}{1 - \frac{\rho_N}{[\rho_N]_{IRR_p^{IPP}}} [1 - (1 - p)(1 - v_m)]}
$$

$$
+ \frac{\rho_{N-M} J_M [1 - (1 - p)(1 - v_m)] \left(\frac{1}{[\rho_N]_{IRR_p^{IPP}}} - \frac{1}{[\rho_M]_{IRR_p^{IPP}}} \right) - \frac{J_M}{(1 + IRR_p^{IPP})^M}}{1 - \frac{\rho_N}{[\rho_N]_{IRR_p^{IPP}}} [1 - (1 - p)(1 - v_m)]}
$$

$$(3.11)$$

As indicated above, formulae (3.1)–(3.10) form a generalization of formulae in (1.1)–(1.3), which are derived from (3.1)–(3.10) by substitution of $v_m = 0$, $J_M = 0$ in the place of v_m and J_M.

The notation of Eqs. (3.1)–(3.10) using series is considered a drawback, due to the necessity to conduct time-consuming and extensive step-by-step process of calculating the consecutive values of the consecutive expressions of the series in the years $t = 1, 2,..., N$, while their summation does not offer the possibility of a quick and straight forward analysis of their variability. For avoiding this inconvenience, one must adopt an assumption that cash flows CF (such as price of energy carriers and environmental charges) are constant in time, just as assumed in [1]. At that point, geometric series describing individual measures *NPV, IRR, DPBP* can be expressed as a sum of their first N terms, and in their compact notations are suitable for analysis [1]. The assumption of the constant cash flow in the time period N does not allow for the possibility of analysis of the value of the heat and electricity market as well as market worth of power stations and heat and power plants supplying the market in case of the fluctuation of such parameters as the price of the energy carriers or environmental charges over time. These problems can be avoided using *NPV, IRR, DPBP* notation (3.1)–(3.3) written over continuous time intervals, i.e. when they are described using integrals. Then, any function of change in the integral value over time can be assumed, for instance any scenarios of the variation in price of energy carriers in time or change in specific tariff for emission of pollutants into natural environment can be considered. Denoting NPV, IRR and DPBP measures in the continuous time as well as methodology of development of "continuous" mathematical models using these measures offer the application of differential calculus in analysis of the variation in their value. In general, continuous models form a foundation of a detailed technical and economic analysis of various investment processes in any branch of economy. This feature gives them an incomparable advantage over discrete notations. Moreover, they offer insight into a range of additional, important information about the characteristics of considered economical processes and phenomena. Consequently, they offer the possibility of assessing the impact of a specific input value on the final result, in addition to which, they provide a solution that is optimal along with a scope of other solutions close to this optimum. They allow the discussion and analysis of results, as well as show the character of their change. This fact is of much importance in practical application. In addition, such continuous models allow to draw the conclusions concerning the economic conditionality of applying particular power engineering technologies on the value of heat and electricity market as well as market worth of power stations and heat and power plants. They enable the analysis of the impact of values such as: demand for heat and electricity, relations and scope of fluctuations in the price of energy carriers etc. on the value of the market and energy sources. To sum up, presented models, finding their application in performing technical and economic analysis of the value of the heat and electricity market as well as market worth of heat and power plants, involve advantages in terms of cognitive and knowledge-expanding characteristics, thus giving them a wide scope of application.

3.2.1   Continuous Mathematical Models of Analysis and Valuation of the Value of the Heat and Electricity Market as Well as Market Worth of Power Stations and Heat and Power Plants Supplying the Market

3.2.1.1   Mathematical Models Excluding Investment on Modernisation of Power Stations and Heat and Power Plants

With regard to continuous time, the discounted measures *NPV, IRR, DPBP* are expressed by the formulae below.

- Net present value *NPV* for *IPP* is

$$NPV^{IPP} = (1 - v_m) \int_0^T [S_A - K_e - F - R - (S_A - K_e - F - A)p]e^{-rt}dt \qquad (3.12)$$

and for the case of an exclusive player in the market the total profit is

$$v_m \int_0^T [S_A - K_e - F - R - (S_A - K_e - F - A)p]e^{-rt}dt \qquad (3.13)$$

where:

v_m relative value of the heat and electricity market ($0 \le v_m \le 1$); this is a share (in per cent) held by the current owner in joined ownership with IPP in a heat or power plant sold to him; v_m is calculated according to the accepted criterion of market valuation; this criterion should be the value of interest rate of IRR^{IPP}

Values of A, F and R in formulae (3.12), (3.13) are functions of the discount rate r, just as in (3.18).

The expression in Eqs. (3.12) and (3.13) [cf. formula (1.12)] in the form

$$Z = S_A - K_e - F - A \tag{3.14}$$

denotes the annual gross profit on operation gained from power station or heat or power plant, and the expression [cf. formula (1.5)]

$$P = (S_A - K_e - F - A)p \tag{3.15}$$

denotes the income tax P on this profit. The lower the tax rate on profit before tax p, the greater the relative value of the market v_m (for the State Treasury the most beneficial condition would involve 100 % tax and total control of the profit; this, however, would be "a road leading to nowhere" as being completely damaging to motivation to operate and do invest).

An increase in operation period in the market T (operation period of a power station or heat or power plant) leads to an increase in profit and the value of the market.

- Internal rate of return IRR^{IPP} (calculated from (3.12) under the assumption that the investment brings no profit, i.e. the income is equal to expenses, hence $NPV = 0$, and the tax $P = 0$ [1]) is expressed in the form

$$\int_0^T [F + A + (S_A - K_e - F - A)(1 - v_m)]e^{-IRR^{IPP}t}dt$$

$$= \int_0^T [F(IRR^{IPP}) + R(IRR^{IPP})]e^{-IRR^{IPP}t}dt \tag{3.16}$$

The terms $F(IRR^{IPP})$ and $R(IRR^{IPP})$ on the right-hand side of the Eq. (3.16) denote the financial cost F and the loan installment R are both functions of the internal return rate IRR^{IPP}, while on the left-hand side these measures along with the depreciation installment A solely constitute functions of the discount rate r. Moreover, the right -hand side of the Eq. (3.16) describes the investment and the interest rate on it, while the sum $F + A$ in the formulae (3.12)–(3.16) represents depreciation including interest $\rho_N J_0$ in Eqs. (3.4)–(3.9).

The value of the total internal return rate on the investment IRR is derived from the Eq. (3.16) for $v_m = 0$. Obviously, the relation $IRR^{IPP} \leq IRR$ must also apply in this case (in case of $IRR^{IPP} = IRR$, the only owner of the power station or heat or power plant is IPP). Otherwise, the market of heat and electricity is completely unattractive to the IPP investor. As mentioned above, IRR is calculated from the formula (3.16) according to its definition, under the assumption that no profit is gained and income tax (formula 3.15) is equal to zero $P = 0$. Therefore, formula

(3.16) gives the gross cash flow. The annual profit $D(1 - v_m)$ gained by the *IPP* investor represents the profit after tax and so *IPP* investor is likely to request that the net calculation of IRR^{IPP} for this profit is made. Thus, taking into account the income tax from Z, the internal return rate IRR_p^{IPP} can be derived on the basis of the relation:

$$\int_0^T [F + A + (S_A - K_e - F - A)(1 - p)(1 - v_m)]e^{-IRR_p^{IPP}t}dt$$

$$= \int_0^T \left[F(IRR_p^{IPP}) + R(IRR_p^{IPP}) \right] e^{-IRR_p^{IPP}t}dt \tag{3.17}$$

In these circumstances, i.e. for the value of IRR_p^{IPP} requested by the investor, the relative market value v_m decreases while the profit gained by the *IPP* increases (formula 3.12).

The formula (3.17) can also be applied to calculate the relative market value v_m adopted by the IPP, the value of the return rate of the investment made on the purchase of the heat or power plant, based on a given annual exploitation cost K_e, annual revenue from operation of heat or power plant S_A, and the value of the investment J.

The difference between the annual revenue S_A (which is relative to the price of goods and services as well as the total production sold in the market) and annual exploitation cost K_e imposes the level of the IRR^{IPP} and IRR_p^{IPP} rates. If the difference between the revenue S_A and the exploitation cost K_e increases, consequently, the values of IRR, IRR^{IPP}, IRR_p^{IPP} increase as well. The relative market value v_m of the goods and service market increases with a constant value of IRR^{IPP}, IRR_p^{IPP}. When $v_m = 100\,\%$ (the only owner of the company is the exclusive player in the market) IRR^{IPP}, IRR_p^{IPP} are equal to the discount rate r and *IPP* is the only creditor of the capital J with the interest rate equal to r.

A decrease in the tax rate p results in an increase in relative market value v_m (a rise in tax p leads to decrease in profit Z and stops investor's motivation to work and invest). An increase in the duration of the long term contracts, i.e. the market exploitation period T (period of operation of the company) also increases the profit and, consequently, the value of the market increases.

- Discounted payback period of the investment expenditure $DPBP^{IPP}$ (determined under the assumption of $NPV = 0$)

$$\int_0^{DPBP^{IPP}} [F + A + (S_A - K_e - F - A)(1 - p)(1 - v_m)]e^{-rt}dt = \int_0^T (F + R)e^{-rt}dt$$

$$\tag{3.18}$$

while the expression [cf. formula (1.13)]

$$D = (S_A - K_e - F - A)(1 - p) \tag{3.19}$$

denotes the annual net profit from operation of a power station or heat or power plant.

Dynamic payback period $DPBP$ of the investment J is determined for $v_m = 0$. Obviously, a relation such that $DPBP < DPBP^{IPP}$ occurs there.

After substituting Eqs. (2.19), (2.22), (2.24)–(2.39) into (3.12), (3.16), (3.17), (3.18) and integrating within given limits (i.e. assuming that J, N_{el}, p, r, t_A, ε_{el}, η_{el}, δ_{serv}, p_{CO_2}, p_{CO}, p_{NO_x}, p_{SO_2}, p_{dust}, a_{el}, a_{coal}, a_{CO_2}, a_{CO}, a_{SO_2}, a_{NO_x}, a_{dust}, b_{CO_2}, u are constant in time), we can obtain the relations (as shown below) for NPV^{IPP}, IRR_p^{IPP}, $DPBP^{IPP}$ and the market worth of the power station and heat and power plant J.

- Net Present Value NPV^{IPP}:

$$
\begin{aligned}
NPV^{IPP} = \Big\{ & N_{el}(1 - \varepsilon_{el})t_A e_{el}^{t=0} \frac{1}{a_{el} - r}[e^{(a_{el}-r)T} - 1] \\
& - \frac{N_{el}t_A}{\eta_{el}}(1 + x_{sw,m,was})e_{coal}^{t=0} \frac{1}{a_{coal} - r}[e^{(a_{coal}-r)T} - 1] \\
& - \frac{N_{el}t_A}{\eta_{el}} p_{CO_2} p_{CO_2}^{t=0} \frac{1}{a_{CO_2} - r}[e^{(a_{CO_2}-r)T} - 1] \\
& - \frac{N_{el}t_A}{\eta_{el}} p_{CO} p_{CO}^{t=0} \frac{1}{a_{CO} - r}[e^{(a_{CO}-r)T} - 1] \\
& - \frac{N_{el}t_A}{\eta_{el}} p_{NO_x} p_{NO_x}^{t=0} \frac{1}{a_{NO_x} - r}[e^{(a_{NO_x}-r)T} - 1] \\
& - \frac{N_{el}t_A}{\eta_{el}} p_{SO_2} p_{SO_2}^{t=0} \frac{1}{a_{SO_2} - r}[e^{(a_{SO_2}-r)T} - 1] \\
& - \frac{N_{el}t_A}{\eta_{el}} p_{dust} p_{dust}^{t=0} \frac{1}{a_{dust} - r}[e^{(a_{dust}-r)T} - 1] \\
& - \frac{N_{el}t_A}{\eta_{el}}(1 - u)p_{CO_2} e_{CO_2}^{t=0} \frac{1}{b_{CO_2} - r}[e^{(b_{CO_2}-r)T} - 1] \\
& - J(1 - e^{-rT})(1 + x_{sal,t,ins})\frac{\delta_{serv}}{r} - J_0(\frac{1 - e^{-rT}}{T} + 1) \Big\}(1 - p)(1 - v_m).
\end{aligned}
\tag{3.20}
$$

Formula for the total NPV is determined from (3.20) for $v_m = 0$.

- Internal Rate of Return for the investor IRR_p^{IPP}:

$$J\frac{(1+r)^{b+1}-1}{(b+1)r}\left[\frac{r}{IRR_p^{IPP}}+\frac{1-e^{-IRR_p^{IPP}T}}{T}\left(\frac{r}{IRR_p^{IPP}}+\frac{1}{IRR_p^{IPP}}-\frac{r}{(IRR_p^{IPP})^2}\right)\right]$$

$$+\left\{N_{el}(1-\varepsilon_{el})t_A e_{el}^{t=0}\frac{1}{a_{el}-IRR_p^{IPP}}\left[e^{(a_{el}-IRR_p^{IPP})T}-1\right]\right.$$

$$-\frac{N_{el}t_A}{\eta_{el}}(1+x_{sw,w,was})e_{coal}^{t=0}\frac{1}{a_{coal}-IRR_p^{IPP}}\left[e^{(a_{coal}-IRR_p^{IPP})T}-1\right]$$

$$-\frac{N_{el}t_A}{\eta_{el}}\rho_{CO_2}p_{CO_2}^{t=0}\frac{1}{a_{CO_2}-IRR_p^{IPP}}\left[e^{(a_{CO_2}-IRR_p^{IPP})T}-1\right]$$

$$-\frac{N_{el}t_A}{\eta_{el}}\rho_{CO}p_{CO}^{t=0}\frac{1}{a_{CO}-IRR_p^{IPP}}\left[e^{(a_{CO}-IRR_p^{IPP})T}-1\right]$$

$$-\frac{N_{el}t_A}{\eta_{el}}\rho_{NO_X}p_{NO_X}^{t=0}\frac{1}{a_{NO_X}-IRR_p^{IPP}}\left[e^{(a_{NO_X}-IRR_p^{IPP})T}-1\right]$$

$$-\frac{N_{el}t_A}{\eta_{el}}\rho_{SO_2}p_{SO_2}^{t=0}\frac{1}{a_{SO_2}-IRR_p^{IPP}}\left[e^{(a_{SO_2}-IRR_p^{IPP})T}-1\right]$$

$$-\frac{N_{el}t_A}{\eta_{el}}\rho_{dust}p_{dust}^{t=0}\frac{1}{a_{dust}-IRR_p^{IPP}}\left[e^{(a_{dust}-IRR_p^{IPP})T}-1\right]$$

$$-\frac{N_{el}t_A}{\eta_{el}}(1-u)\rho_{CO_2}e_{CO_2}^{t=0}\frac{1}{b_{CO_2}-IRR_p^{IPP}}\left[e^{(b_{CO_2}-IRR_p^{IPP})T}-1\right]$$

$$-J(1+x_{sal,t,ins})\frac{\delta_{serv}}{IRR_p^{IPP}}(1-e^{-IRR_p^{IPP}T})$$

$$\left.-J\frac{(1+r)^{b+1}-1}{(b+1)r}\left[\frac{r}{IRR_p^{IPP}}+\frac{1-e^{-IRR_p^{IPP}T}}{T}\left(\frac{r}{IRR_p^{IPP}}+\frac{1}{IRR_p^{IPP}}-\frac{r}{(IRR_p^{IPP})^2}\right)\right]\right\}$$

$$\times(1-p)(1-v_m)=J\frac{(1+IRR_p^{IPP})^{b+1}-1}{(b+1)IRR_p^{IPP}}\left(1+\frac{1-e^{-IRR_p^{IPP}T}}{T}\right).$$

$$(3.21)$$

The calculation of IRR_p^{IPP} requires the successive application of an approximation method.

The formula for the total internal return rate IRR is derived from (3.21) for $p = 0$ and $v_m = 0$, the formula for IRR^{IPP}, i.e. for $p = 0$.

From the Eq. (3.21), as mentioned above, one can calculate the relative market value v_m for given investment J and for the assumed by the IPP investor interest rate IRR_p^{IPP} of the capital invested in the purchase of a power station or heat and power plant.

$$v_m = \left\{ J \frac{(1+r)^{b+1}-1}{(b+1)r} \left[\frac{r}{IRR_p^{IPP}} + \frac{1-e^{-IRR_p^{IPP}T}}{T} \left(\frac{r}{IRR_p^{IPP}} + \frac{1}{IRR_p^{IPP}} - \frac{r}{(IRR_p^{IPP})^2} \right) \right] \right.$$
$$\left. - J \frac{(1+IRR_p^{IPP})^{b+1}-1}{(b+1)IRR_p^{IPP}} \left(1 + \frac{1-e^{-IRR_p^{IPP}T}}{T} \right) \right\} \frac{1}{B} + 1$$

$$(3.22)$$

where

$$B = \left\{ N_{el}(1-\varepsilon_{el}) t_A e_{el}^{t=0} \frac{1}{a_{el}-IRR_p^{IPP}} [e^{(a_{el}-IRR_p^{IPP})T} - 1] \right.$$

$$- \frac{N_{el}t_A}{\eta_{el}} (1+x_{sw,m,was}) e_{coal}^{t=0} \frac{1}{a_{coal}-IRR_p^{IPP}} [e^{(a_{coal}-IRR_p^{IPP})T} - 1]$$

$$- \frac{N_{el}t_A}{\eta_{el}} \rho_{CO_2} p_{CO_2}^{t=0} \frac{1}{a_{CO_2}-IRR_p^{IPP}} [e^{(a_{CO_2}-IRR_p^{IPP})T} - 1]$$

$$- \frac{N_{el}t_A}{\eta_{el}} \rho_{CO} p_{CO}^{t=0} \frac{1}{a_{CO}-IRR_p^{IPP}} [e^{(a_{CO}-IRR_p^{IPP})T} - 1]$$

$$- \frac{N_{el}t_A}{\eta_{el}} \rho_{NO_X} p_{NO_X}^{t=0} \frac{1}{a_{NO_X}-IRR_p^{IPP}} [e^{(a_{NO_X}-IRR_p^{IPP})T} - 1]$$

$$- \frac{N_{el}t_A}{\eta_{el}} \rho_{SO_2} p_{SO_2}^{t=0} \frac{1}{a_{SO_2}-IRR_p^{IPP}} [e^{(a_{SU_2}-IRR_p^{IPP})T} - 1]$$

$$- \frac{N_{el}t_A}{\eta_{el}} \rho_{dust} p_{dust}^{t=0} \frac{1}{a_{dust}-IRR_p^{IPP}} [e^{(a_{dust}-IRR_p^{IPP})T} - 1]$$

$$- \frac{N_{el}t_A}{\eta_{el}} (1-u) \rho_{CO_2} e_{CO_2}^{t=0} \frac{1}{b_{CO_2}-IRR_p^{IPP}} [e^{(b_{CO_2}-IRR_p^{IPP})T} - 1]$$

$$- J(1+x_{sal,t,ins}) \frac{\delta_{serv}}{IRR_p^{IPP}} (1 - e^{-IRR_p^{IPP}T})$$

$$\left. - J \frac{(1+r)^{b+1}-1}{(b+1)r} \left[\frac{r}{IRR_p^{IPP}} + \frac{1-e^{-IRR_p^{IPP}T}}{T} \left(\frac{r}{IRR_p^{IPP}} + \frac{1}{IRR_p^{IPP}} - \frac{r}{(IRR_p^{IPP})^2} \right) \right] \right\} (1-p)$$

$$(3.23)$$

- Market worth of a power plant J

We can also determine the market worth of a power station J from the formula in (3.21). Then, it is a function of relative market value v_m, interest rates r and IRR_p^{IPP}, prices of the energy carriers and environmental charges. In addition, if the income from marketing of heat and chemical energy use of the fuel for its production were included, formula (3.21) would describe the market worth of a heat and power plant.

$$
\begin{aligned}
J = \Bigg\{ & N_{el}(1 - \varepsilon_{el}) t_A e_{el}^{t=0} \frac{1}{a_{el} - IRR_p^{IPP}} [e^{(a_{el} - IRR_p^{IPP})T} - 1] \\
& - \frac{N_{el} t_A}{\eta_{el}} (1 + x_{sw,m,was}) e_{coal}^{t=0} \frac{1}{a_{coal} - IRR_p^{IPP}} [e^{(a_{coal} - IRR_p^{IPP})T} - 1] \\
& - \frac{N_{el} t_A}{\eta_{el}} \rho_{CO_2} p_{CO_2}^{t=0} \frac{1}{a_{CO_2} - IRR_p^{IPP}} [e^{(a_{CO_2} - IRR_p^{IPP})T} - 1] \\
& - \frac{N_{el} t_A}{\eta_{el}} \rho_{CO} p_{CO}^{t=0} \frac{1}{a_{CO} - IRR_p^{IPP}} [e^{(a_{CO} - IRR_p^{IPP})T} - 1] \\
& - \frac{N_{el} t_A}{\eta_{el}} \rho_{NO_X} p_{NO_X}^{t=0} \frac{1}{a_{NO_X} - IRR_p^{IPP}} [e^{(a_{NO_X} - IRR_p^{IPP})T} - 1] \\
& - \frac{N_{el} t_A}{\eta_{el}} \rho_{SO_2} p_{SO_2}^{t=0} \frac{1}{a_{SO_2} - IRR_p^{IPP}} [e^{(a_{SO_2} - IRR_p^{IPP})T} - 1] \\
& - \frac{N_{el} t_A}{\eta_{el}} \rho_{dust} p_{dust}^{t=0} \frac{1}{a_{dust} - IRR_p^{IPP}} [e^{(a_{dust} - IRR_p^{IPP})T} - 1] \\
& - \frac{N_{el} t_A}{\eta_{el}} (1 - u) \rho_{CO_2} e_{CO_2}^{t=0} \frac{1}{b_{CO_2} - IRR_p^{IPP}} [e^{(b_{CO_2} - IRR_p^{IPP})T} - 1] \Bigg\} \frac{(1-p)(1-v_m)}{C}
\end{aligned}
$$

$$(3.24)$$

where

$$
\begin{aligned}
C = & \frac{(1 + IRR_p^{IPP})^{b+1} - 1}{(b+1)IRR_p^{IPP}} \left(1 + \frac{1 - e^{-IRR_p^{IPP} T}}{T} \right) \\
& + \frac{(1+r)^{b+1} - 1}{(b+1)r} \left[\frac{r}{IRR_p^{IPP}} + \frac{1 - e^{-IRR_p^{IPP} T}}{T} \left(\frac{r}{IRR_p^{IPP}} + \frac{1}{IRR_p^{IPP}} - \frac{r}{(IRR_p^{IPP})^2} \right) \right] \\
& \times [(1-p)(1-v_m) - 1] \\
& + (1 + x_{sal,t,ins}) \frac{\delta_{serv}}{IRR_p^{IPP}} (1 - e^{-r IRR_p^{IPP}})(1-p)(1-v_m).
\end{aligned}
$$

$$(3.25)$$

The price of J is very sensitive to the variation in the rate of interest IRR_p^{IPP}. An increase in IRR_p^{IPP} by 1 or 2 % could result in a subsequent alteration of the price of J by up to a couple of dozen or so per cent.

The formula (3.24) can be applied to determine the economically justified investment J in building a power station or heat and power plant with given relations between prices of energy carriers (this is an opposite issue to establishing the economically justified prices of energy carriers with real investment J resulting from actual expenses on construction, prices of the building material, price of equipment etc.).

The total expenditure J borne by the IPP purchasing an energy source from the previous owner (e.g. the State Treasury) with a desired value of IRR_p^{IPP} and the relative market value v_m demanded by the previous owner should be at a maximum equal to the investment expenditure on a completely newly constructed energy source. Construction of a new energy source will make sense only when there is a demand for the heat and electricity produced by the IPP.

- Dynamic Pay-Back Period $DPBP^{IPP}$:

$$J_0\left[1+\frac{1}{T}-e^{-rDPBP^{IPP}}\left(1+\frac{1}{T}-\frac{DPBP^{IPP}}{T}\right)\right]$$

$$+\left\{N_{el}(1-\varepsilon_{el})t_A e_{el}^{t=0}\frac{1}{a_{el}-r}[e^{(a_{el}-r)DPBP^{IPP}}-1]-\frac{N_{el}t_A}{\eta_{el}}(1+x_{sw,m,was})e_{coal}^{t=0}\frac{1}{a_{coal}-r}[e^{(a_{coal}-r)DPBP^{IPP}}-1]\right.$$

$$-\frac{N_{el}t_A}{\eta_{el}}\rho_{CO_2}p_{CO_2}^{t=0}\frac{1}{a_{CO_2}-r}[e^{(a_{CO_2}-r)DPBP^{IPP}}-1]-\frac{N_{el}t_A}{\eta_{el}}\rho_{CO}p_{CO}^{t=0}\frac{1}{a_{CO}-r}[e^{(a_{CO}-r)DPBP^{IPP}}-1]$$

$$-\frac{N_{el}t_A}{\eta_{el}}\rho_{NO_X}p_{NO_X}^{t=0}\frac{1}{a_{NO_X}-r}[e^{(a_{NO_X}-r)DPBP^{IPP}}-1]-\frac{N_{el}t_A}{\eta_{el}}\rho_{SO_2}p_{SO_2}^{t=0}\frac{1}{a_{SO_2}-r}[e^{(a_{SO_2}-r)DPBP^{IPP}}-1]$$

$$-\frac{N_{el}t_A}{\eta_{el}}\rho_{dust}p_{dust}^{t=0}\frac{1}{a_{dust}-r}[e^{(a_{dust}-r)DPBP^{IPP}}-1]-\frac{N_{el}t_A}{\eta_{el}}(1-u)\rho_{CO_2}e_{CO_2}^{t=0}\frac{1}{b_{CO_2}-r}[e^{(b_{CO_2}-r)DPBP^{IPP}}-1]$$

$$\left.-J(1-e^{-rDPBP^{IPP}})(1+x_{sal,t,ins})\frac{\delta_{serv}}{r}-J_0\left[1+\frac{1}{T}-e^{-rDPBP^{IPP}}\left(1+\frac{1}{T}-\frac{DPBP^{IPP}}{T}\right)\right]\right\}(1-p)(1-v_m)$$

$$=J_0(1+\frac{1-e^{-rT}}{T}).$$

$$(3.26)$$

The formula with regard to the dynamic payback period $DPBP$ is derived from (3.26) for $v_m = 0$.

The calculation of the value of $DPBP$ requires an application of a successive approximation method.

3.2.1.2 Mathematical Models Including Investment in Modernization of a Power Station or Heat and Power Plant

The above mathematical models do not account for the investment expenditure J_M needed on the modernization of the energy sources, which leads to a decrease in the value of the market and their market worth. The question is—to what extent does this happen? This question can be answered using analysis applying the models presented below.

- Net Present Value NPV^{IPP} gained by the owner of an energy source:

$$NPV^{IPP} = \int_0^{t_1} [F + A + (S - K_e - F - A)(1 - p)(1 - v_m)]e^{-rt}dt$$

$$+ \int_{t_1}^{t_2} [F + A + F^M + A^M + (S^M - K_e^M - F - A - F^M - A^M)(1 - p)(1 - v_m)]e^{-rt}dt$$

$$+ \int_{t_2}^{T} [F + A + F^M + A^M + (S^{mod} - K_e^{mod} - F - A - F^M - A^M)(1 - p)(1 - v_m)]e^{-rt}dt$$

$$- \int_0^{T} (F + R)e^{-rt}dt - \int_{t_1}^{T} (F^M + R^M)e^{-rt}dt$$

$$= NPV(1 - v_m) = \left[\int_0^{t_1} (S - K_e - F - A)(1 - p)e^{-rt}dt \right.$$

$$+ \int_{t_1}^{t_2} (S^M - K_e^M - F - A - F^M - A^M)(1 - p)e^{-rt}dt$$

$$+ \left. \int_{t_2}^{T} (S^{mod} - K_e^{mod} - F - A - F^M - A^M)(1 - p)e^{-rt}dt \right](1 - v_m)$$

$$(3.27)$$

with

$$A^M = R^M = \frac{J_M}{T - t_1} \qquad (3.28)$$

and time intervals $\langle 0, t_1 \rangle$, $\langle t_1, t_2 \rangle$, $\langle t_2, T \rangle$ represent the exploitation of the energy source before, during and after their modernization, respectively (cf. Fig. 3.1).

- The Internal Rate of Return IRR_p^{IPP} of the investment expenditure J, J_M, associated with the purchase and modernisation of an energy source

$$\int_0^{t_1} [F(r) + A(r)]e^{-IRR_p^{IPP}t}dt + \left\{\int_0^{t_1} [S - K_e - F(r) - A(r)]e^{-IRR_p^{IPP}t}dt\right\}(1-p)(1-v_m)$$

$$+ \int_{t_1}^{t_2} [F(r) + A(r) + F^M(r) + A^M(r)]e^{-IRR_p^{IPP}t}dt + \left\{\int_{t_1}^{t_2} [S^M - K_e^M\right.$$

$$\left. - F(r) - A(r) - F^M(r) - A^M(r)]e^{-IRR_p^{IPP}t}dt\right\}(1-p)(1-v_m)$$

$$+ \int_{t_2}^{T} [F(r) + A(r) + F^M(r) + A^M(r)]e^{-IRR_p^{IPP}t}dt + \left\{\int_{t_2}^{T} [S^{mod} - K_e^{mod}\right.$$

$$\left. - F(r) - A(r) - F^M(r) - A^M(r)]e^{-IRR_p^{IPP}t}dt\right\}(1-p)(1-v_m)$$

$$= \int_0^{T} [F(IRR_p^{IPP}) + R(IRR_p^{IPP})]e^{-IRR_p^{IPP}t}dt + \int_{t_1}^{T} [F^M(IRR_p^{IPP}) + R^M(IRR_p^{IPP})]e^{-IRR_p^{IPP}t}dt$$

$$(3.29)$$

The formula in (3.29) is used to determine the market worth of an energy source J. It is a function of relative market value v_m, interest rates r and IRR_p^{IPP}, prices of energy carriers, environmental charges and investment J_M in the modernization of an energy source.

- Discounted Pay-Back Period $DPBP^{IPP}$ of an investment J, J_M, in the purchase and modernisation of an energy source:

$$\int_0^{t_1} [F + A + (S - K_e - F - A)(1-p)(1-v_m)]e^{-rt}dt$$

$$+ \int_{t_1}^{t_2} [F + A + F^M + A^M + (S^M - K_e^M - F - A - F^M - A^M)(1-p)(1-v_m)]e^{-rt}dt$$

$$+ \int_{t_2}^{DPBP^{IPP}} [F + A + F^M + A^M + (S^{mod} - K_e^{mod} - F - A - F^M - A^M)(1-p)(1-v_m)]e^{-rt}dt$$

$$= \int_0^{T} (F + R)e^{-rt}dt + \int_{t_1}^{T} (F^M + R^M)e^{-rt}dt$$

$$(3.30)$$

Substituting Eqs. (2.19), (2.22), (2.24)–(2.39), (3.28) into (3.27), (3.29), (3.30) and integration of the final result, we arrive at the final formulae for NPV^{IPP}, IRR_p^{IPP}, $DPBP^{IPP}$.

- Net Present Value NPV^{IPP}

$$
\begin{aligned}
NPV^{IPP} = &\left\{ \left\{ N_{el}(1 - \varepsilon_{el}) t_A e_{el}^{t=0} \frac{1}{a_{el} - r} \left[e^{(a_{el}-r)t_1} - 1 \right] \right. \right. \\
&- \frac{N_{el} t_A}{\eta_{el}} (1 + x_{sw,m,was}) e_{coal}^{t=0} \frac{1}{a_{coal} - r} \left[e^{(a_{coal}-r)t_1} - 1 \right] \\
&- \frac{N_{el} t_A}{\eta_{el}} \rho_{CO_2} p_{CO_2}^{t=0} \frac{1}{a_{CO_2} - r} \left[e^{(a_{CO_2}-r)t_1} - 1 \right] \\
&- \frac{N_{el} t_A}{\eta_{el}} \rho_{CO} p_{CO}^{t=0} \frac{1}{a_{CO} - r} \left[e^{(a_{CO}-r)t_1} - 1 \right] \\
&- \frac{N_{el} t_A}{\eta_{el}} \rho_{NO_X} p_{NO_X}^{t=0} \frac{1}{a_{NO_X} - r} \left[e^{(a_{NO_X}-r)t_1} - 1 \right] \\
&- \frac{N_{el} t_A}{\eta_{el}} \rho_{SO_2} p_{SO_2}^{t=0} \frac{1}{a_{SO_2} - r} \left[e^{(a_{SO_2}-r)t_1} - 1 \right] \\
&- \frac{N_{el} t_A}{\eta_{el}} \rho_{dust} p_{dust}^{t=0} \frac{1}{a_{dust} - r} \left[e^{(a_{dust}-r)t_1} - 1 \right] \\
&- \frac{N_{el} t_A}{\eta_{el}} (1 - u) \rho_{CO_2} e_{CO_2}^{t=0} \frac{1}{b_{CO_2} - r} \left[e^{(b_{CO_2}-r)t_1} - 1 \right] \\
&\left. - J(1 + x_{sal,t,ins}) \frac{\delta_{serv}}{r} (1 - e^{-rt_1}) - J_0 \left[1 + \frac{1}{T} - \left(1 + \frac{1}{T} - \frac{t_1}{T} \right) e^{-rt_1} \right] \right\} \\
&+ \left\{ N_{el}^M (1 - \varepsilon_{el}^M) t_A^M e_{el}^{M,t=t_1} \frac{1}{a_{el}^M - r} \left[e^{(a_{el}^M - r)t_2} - e^{(a_{el}^M - r)t_1} \right] \right. \\
&- \frac{N_{el}^M t_A^M}{\eta_{el}^M} (1 + x_{sw,m,was}) e_{coal}^{M,t=t_1} \frac{1}{a_{coal}^M - r} \left[e^{(a_{coal}^M - r)t_2} - e^{(a_{coal}^M - r)t_1} \right] \\
&- \frac{N_{el}^M t_A^M}{\eta_{el}^M} \rho_{CO_2} p_{CO_2}^{M,t=t_1} \frac{1}{a_{CO_2}^M - r} \left[e^{\left(a_{CO_2}^M - r\right)t_2} - e^{\left(a_{CO_2}^M - r\right)t_1} \right] \\
&- \frac{N_{el}^M t_A^M}{\eta_{el}^M} \rho_{CO} p_{CO}^{M,t=t_1} \frac{1}{a_{CO}^M - r} \left[e^{(a_{CO}^M - r)t_2} - e^{(a_{CO}^M - r)t_1} \right] \\
&- \frac{N_{el}^M t_A^M}{\eta_{el}^M} \rho_{NO_X} p_{NO_X}^{M,t=t_1} \frac{1}{a_{NO_X}^M - r} \left[e^{\left(a_{NO_X}^M - r\right)t_2} - e^{\left(a_{NO_X}^M - r\right)t_1} \right] \\
&- \frac{N_{el}^M t_A^M}{\eta_{el}^M} \rho_{SO_2} p_{SO_2}^{M,t=t_1} \frac{1}{a_{SO_2}^M - r} \left[e^{\left(a_{SO_2}^M - r\right)t_2} - e^{\left(a_{SO_2}^M - r\right)t_1} \right] \\
&- \frac{N_{el}^M t_A^M}{\eta_{el}^M} \rho_{dust} p_{dust}^{M,t=t_1} \frac{1}{a_{dust}^M - r} \left[e^{(a_{dust}^M - r)t_1} - e^{(a_{dust}^M - r)t_1} \right] \\
&\left. \left. - \frac{N_{el}^M t_A^M}{\eta_{el}^M} (1 - u^M) \rho_{CO_2} e_{CO_2}^{M,t=t_1} \frac{1}{b_{CO_2}^M - r} \left[e^{\left(b_{CO_2}^M - r\right)t_2} - e^{\left(b_{CO_2}^M - r\right)t_1} \right] \right. \right.
\end{aligned}
$$

$$(3.31)$$

- Internal Rate of Return IRR_p^{IPP}

$$J\frac{(1+r)^{b+1}-1}{(b+1)r}\left\{\left(\frac{r}{TIRR_p^{IPP}}+\frac{r}{IRR_p^{IPP}}+\frac{1}{TIRR_p^{IPP}}\right)\left(1-e^{-IRR_p^{IPP}t_1}\right)\right.$$

$$\left.-\frac{r}{T(IRR_p^{IPP})^2}\left[1-e^{-IRR_p^{IPP}t_1}\left(IRR_p^{IPP}t_1+1\right)\right]\right\}$$

$$+\left\{N_{el}(1-\varepsilon_{el})t_A e_{el}^{t=0}\frac{1}{a_{el}-IRR_p^{IPP}}[e^{(a_{el}-IRR_p^{IPP})t_1}-1]\right.$$

$$-\frac{N_{el}t_A}{\eta_{el}}(1+x_{sw,m,was})e_{coal}^{t=0}\frac{1}{a_{coal}-IRR_p^{IPP}}[e^{(a_{coal}-IRR_p^{IPP})t_1}-1]$$

$$-\frac{N_{el}t_A}{\eta_{el}}\rho_{CO_2}p_{CO_2}^{t=0}\frac{1}{a_{CO_2}-IRR_p^{IPP}}[e^{(a_{CO_2}-IRR_p^{IPP})t_1}-1]$$

$$-\frac{N_{el}t_A}{\eta_{el}}\rho_{CO}p_{CO}^{t=0}\frac{1}{a_{CO}-IRR_p^{IPP}}[e^{(a_{CO}-IRR_p^{IPP})t_1}-1]$$

$$-\frac{N_{el}t_A}{\eta_{el}}\rho_{NO_x}p_{NO_x}^{t=0}\frac{1}{a_{NO_x}-IRR_p^{IPP}}[e^{(a_{NO_x}-IRR_p^{IPP})t_1}-1]$$

$$-\frac{N_{el}t_A}{\eta_{el}}\rho_{SO_2}p_{SO_2}^{t=0}\frac{1}{a_{SO_2}-IRR_p^{IPP}}[e^{(a_{SO_2}-IRR_p^{IPP})t_1}-1]$$

$$-\frac{N_{el}t_A}{\eta_{el}}\rho_{dust}p_{dust}^{t=0}\frac{1}{a_{dust}-IRR_p^{IPP}}[e^{(a_{dust}-IRR_p^{IPP})t_1}-1]$$

$$-\frac{N_{el}t_A}{\eta_{el}}(1-u)\rho_{CO_2}e_{CO_2}^{t=0}\frac{1}{b_{CO_2}-IRR_p^{IPP}}[e^{(b_{CO_2}-IRR_p^{IPP})t_1}-1]$$

$$-J(1+x_{sal,t,ins})\frac{\delta_{serv}}{IRR_p^{IPP}}(1-e^{-IRR_p^{IPP}t_1})$$

$$-J\frac{(1+r)^{b+1}-1}{(b+1)r}\left\{\left(\frac{r}{TIRR_p^{IPP}}+\frac{r}{IRR_p^{IPP}}+\frac{1}{TIRR_p^{IPP}}\right)\left(1-e^{-IRR_p^{IPP}t_1}\right)\right.$$

$$\left.\left.-\frac{r}{T(IRR_p^{IPP})^2}\left[1-e^{-IRR_p^{IPP}t_1}\left(IRR_p^{IPP}t_1+1\right)\right]\right\}\right\}(1-p)(1-v_m)$$

$$+J\frac{(1+r)^{b+1}-1}{(b+1)r}\left\{\left(\frac{r}{TIRR_p^{IPP}}+\frac{r}{IRR_p^{IPP}}+\frac{1}{TIRR_p^{IPP}}\right)\left(e^{-IRR_p^{IPP}t_1}-e^{-IRR_p^{IPP}t_2}\right)\right.$$

$$\left.-\frac{r}{T(IRR_p^{IPP})^2}\left[e^{-IRR_p^{IPP}t_1}\left(IRR_p^{IPP}t_1+1\right)-e^{-IRR_p^{IPP}t_2}\left(IRR_p^{IPP}t_2+1\right)\right]\right\}$$

$$+J_M\left\{\left[\frac{r}{(T-t_1)IRR_p^{IPP}}+\frac{r}{IRR_p^{IPP}}+\frac{1}{(T-t_1)IRR_p^{IPP}}\right]\left(e^{-IRR_p^{IPP}t_1}-e^{-IRR_p^{IPP}t_2}\right)\right.$$

$$- \frac{r}{(T - t_1)(IRR_p^{IPP})^2} \left[e^{-IRR_p^{IPP} t_1} \left(IRR_p^{IPP} t_1 + 1 \right) - e^{-IRR_p^{IPP} t_2} \left(IRR_p^{IPP} t_2 + 1 \right) \right] \Bigg\}$$

$$+ \Bigg\{ N_{el}^M \left(1 - \varepsilon_{el}^M \right) t_A^M e_{el}^{M,t=t_1} \frac{1}{a_{el}^M - IRR_p^{IPP}} \left[e^{\left(a_{el}^M - IRR_p^{IPP} \right) t_2} - e^{\left(a_{el}^M - IRR_p^{IPP} \right) t_1} \right]$$

$$- \frac{N_{el}^M t_A^M}{\eta_{el}^M} e_{coal}^{M,t=t_1} \left(1 + x_{sw,m,was} \right) \frac{1}{a_{coal}^M - IRR_p^{IPP}} \left[e^{\left(a_{coal}^M - IRR_p^{IPP} \right) t_2} - e^{\left(a_{coal}^M - IRR_p^{IPP} \right) t_1} \right]$$

$$- \frac{N_{el}^M t_A^M}{\eta_{el}^M} \rho_{CO_2} p_{CO_2}^{M,t=t_1} \frac{1}{a_{CO_2}^M - IRR_p^{IPP}} \left[e^{\left(a_{CO_2}^M - IRR_p^{IPP} \right) t_2} - e^{\left(a_{CO_2}^M - IRR_p^{IPP} \right) t_1} \right]$$

$$- \frac{N_{el}^M t_A^M}{\eta_{el}^M} \rho_{CO} p_{CO}^{M,t=t_1} \frac{1}{a_{CO}^M - IRR_p^{IPP}} \left[e^{\left(a_{CO}^M - IRR_p^{IPP} \right) t_2} - e^{\left(a_{CO}^M - IRR_p^{IPP} \right) t_1} \right]$$

$$- \frac{N_{el}^M t_A^M}{\eta_{el}^M} \rho_{NO_x} p_{NO_x}^{M,t=t_1} \frac{1}{a_{NO_x}^M - IRR_p^{IPP}} \left[e^{\left(a_{NO_x}^M - IRR_p^{IPP} \right) t_2} - e^{\left(a_{NO_x}^M - IRR_p^{IPP} \right) t_1} \right]$$

$$- \frac{N_{el}^M t_A^M}{\eta_{el}^M} \rho_{SO_2} p_{SO_2}^{M,t=t_1} \frac{1}{a_{SO_2}^M - IRR_p^{IPP}} \left[e^{\left(a_{SO_2}^M - IRR_p^{IPP} \right) t_2} - e^{\left(a_{SO_2}^M - IRR_p^{IPP} \right) t_1} \right]$$

$$- \frac{N_{el}^M t_A^M}{\eta_{el}^M} \rho_{dust} p_{dust}^{M,t=t_1} \frac{1}{a_{dust}^M - IRR_p^{IPP}} \left[e^{\left(a_{dust}^M - IRR_p^{IPP} \right) t_2} - e^{\left(a_{dust}^M - IRR_p^{IPP} \right) t_1} \right]$$

$$- \frac{N_{el}^M t_A^M}{\eta_{el}^M} \left(1 - u^M \right) \rho_{CO_2} e_{CO_2}^{M,t=t_1} \frac{1}{b_{CO_2}^M - IRR_p^{IPP}} \left[e^{\left(b_{CO_2}^M - IRR_p^{IPP} \right) t_2} - e^{\left(b_{CO_2}^M - IRR_p^{IPP} \right) t_1} \right]$$

$$- \frac{(J + J_M)(1 + x_{sal,t,ins}) \delta_{serv}^M}{IRR_p^{IPP}} \left(e^{-IRR_p^{IPP} t_1} - e^{-IRR_p^{IPP} t_2} \right)$$

$$- J \frac{(1 + r)^{b+1} - 1}{(b+1) r} \Bigg\{ \left(\frac{r}{T \, IRR_p^{IPP}} + \frac{r}{IRR_p^{IPP}} + \frac{1}{T \, IRR_p^{IPP}} \right) \left(e^{-IRR_p^{IPP} t_1} - e^{-IRR_p^{IPP} t_2} \right)$$

$$- \frac{r}{T \left(IRR_p^{IPP} \right)^2} \left[e^{-IRR_p^{IPP} t_1} \left(IRR_p^{IPP} t_1 + 1 \right) - e^{-IRR_p^{IPP} t_2} \left(IRR_p^{IPP} t_2 + 1 \right) \right] \Bigg\}$$

$$- J_M \Bigg\{ \left[\frac{r}{(T - t_1) IRR_p^{IPP}} + \frac{r}{IRR_p^{IPP}} + \frac{1}{(T - t_1) IRR_p^{IPP}} \right] \left(e^{-IRR_p^{IPP} t_1} - e^{-IRR_p^{IPP} t_2} \right)$$

$$- \frac{r}{(T - t_1) \left(IRR_p^{IPP} \right)^2} \left[e^{-IRR_p^{IPP} t_1} \left(IRR_p^{IPP} t_1 + 1 \right) - e^{-IRR_p^{IPP} t_2} \left(IRR_p^{IPP} t_2 + 1 \right) \right] \Bigg\} \Bigg\} (1 - p)(1 - v_m)$$

$$+ J \frac{(1 + r)^{b+1} - 1}{(b+1) r} \Bigg\{ \left(\frac{r}{T \, IRR_p^{IPP}} + \frac{r}{IRR_p^{IPP}} + \frac{1}{T \, IRR_p^{IPP}} \right) \left(e^{-IRR_p^{IPP} t_2} - e^{-IRR_p^{IPP} T} \right)$$

$$- \frac{r}{T\left(IRR_p^{IPP}\right)^2}\left[e^{-IRR_p^{IPP}t_2}\left(IRR_p^{IPP}t_2+1\right)-e^{-IRR_p^{IPP}T}\left(IRR_p^{IPP}T+1\right)\right]\Big\}$$

$$+J_M\left\{\left[\frac{r}{(T-t_1)\,IRR_p^{IPP}}+\frac{r}{IRR_p^{IPP}}+\frac{1}{(T-t_1)\,IRR_p^{IPP}}\right]\left(e^{-IRR_p^{IPP}t_2}-e^{-IRR_p^{IPP}T}\right)\right.$$

$$\left.-\frac{r}{(T-t_1)\left(IRR_p^{IPP}\right)^2}\left[e^{-IRR_p^{IPP}t_2}\left(IRR_p^{IPP}t_2+1\right)-e^{-IRR_p^{IPP}T}\left(IRR_p^{IPP}T+1\right)\right]\right\}$$

$$+\left\{N_{el}^{mod}\left(1-\varepsilon_{el}^{mod}\right)t_A^{mod}e_{el}^{mod,t=t_2}\frac{1}{a_{el}^{mod}-IRR_p^{IPP}}\left[e^{\left(a_{el}^{mod}-IRR_p^{IPP}\right)T}-e^{\left(a_{el}^{mod}-IRR_p^{IPP}\right)t_2}\right]\right.$$

$$-\frac{N_{el}^{mod}t_A^{mod}}{\eta_{el}^{mod}}\left(1+x_{sw,m,was}\right)e_{coal}^{mod,t=t_2}\frac{1}{a_{coal}^{mod}-IRR_p^{IPP}}\left[e^{\left(a_{coal}^{mod}-IRR_p^{IPP}\right)T}-e^{\left(a_{coal}^{mod}-IRR_p^{IPP}\right)t_2}\right]$$

$$-\frac{N_{el}^{mod}t_A^{mod}}{\eta_{el}^{mod}}\rho_{CO_2}p_{CO_2}^{mod,t=t_2}\frac{1}{a_{CO_2}^{mod}-IRR_p^{IPP}}\left[e^{\left(a_{CO_2}^{mod}-IRR_p^{IPP}\right)T}-e^{\left(a_{CO_2}^{mod}-IRR_p^{IPP}\right)t_2}\right]$$

$$-\frac{N_{el}^{mod}t_A^{mod}}{\eta_{el}^{mod}}\rho_{CO}p_{CO}^{mod,t=t_2}\frac{1}{a_{CO}^{mod}-IRR_p^{IPP}}\left[e^{\left(a_{CO}^{mod}-IRR_p^{IPP}\right)T}-e^{\left(a_{CO}^{mod}-IRR_p^{IPP}\right)t_2}\right]$$

$$-\frac{N_{el}^{mod}t_A^{mod}}{\eta_{el}^{mod}}\rho_{NO_X}p_{NO_X}^{mod,t=t_2}\frac{1}{a_{NO_X}^{mod}-IRR_p^{IPP}}\left[e^{\left(a_{NO_X}^{mod}-IRR_p^{IPP}\right)T}-e^{\left(a_{NO_X}^{mod}-IRR_p^{IPP}\right)t_2}\right]$$

$$-\frac{N_{el}^{mod}t_A^{mod}}{\eta_{el}^{mod}}\rho_{SO_2}p_{SO_2}^{mod,t=t_2}\frac{1}{a_{SO_2}^{mod}-IRR_p^{IPP}}\left[e^{\left(a_{SO_2}^{mod}-IRR_p^{IPP}\right)T}-e^{\left(a_{SO_2}^{mod}-IRR_p^{IPP}\right)t_2}\right]$$

$$-\frac{N_{el}^{mod}t_A^{mod}}{\eta_{el}^{mod}}\rho_{dust}p_{dust}^{mod,t=t_2}\frac{1}{a_{dust}^{mod}-IRR_p^{IPP}}\left[e^{\left(a_{dust}^{mod}-IRR_p^{IPP}\right)T}-e^{\left(a_{dust}^{mod}-IRR_p^{IPP}\right)t_2}\right]$$

$$-\frac{N_{el}^{mod}t_A^{mod}}{\eta_{el}^{mod}}\left(1-u^{mod}\right)\rho_{CO_2}e_{CO_2}^{mod,t=t_2}\frac{1}{b_{CO_2}^{mod}-IRR_p^{IPP}}\left[e^{\left(b_{CO_2}^{mod}-IRR_p^{IPP}\right)T}-e^{\left(b_{CO_2}^{mod}-IRR_p^{IPP}\right)t_2}\right]$$

$$-\frac{(J+J_M)(1+x_{sal,t,ins})\delta_{serv}^{mod}}{IRR_p^{IPP}}\left(e^{-IRR_p^{IPP}t_2}-e^{-IRR_p^{IPP}T}\right)$$

$$-J\frac{(1+r)^{b+1}-1}{(b+1)r}\left\{\left(\frac{r}{T\,IRR_p^{IPP}}+\frac{r}{IRR_p^{IPP}}+\frac{1}{T\,IRR_p^{IPP}}\right)\left(e^{-IRR_p^{IPP}t_2}-e^{-IRR_p^{IPP}T}\right)\right.$$

$$\left.-\frac{r}{T\left(IRR_p^{IPP}\right)^2}\left[e^{-IRR_p^{IPP}t_2}\left(IRR_p^{IPP}t_2+1\right)-e^{-IRR_p^{IPP}T}\left(IRR_p^{IPP}T+1\right)\right]\right\}$$

$$-J_M\left\{\left[\frac{r}{(T-t_1)\,IRR_p^{IPP}}+\frac{r}{IRR_p^{IPP}}+\frac{1}{(T-t_1)\,IRR_p^{IPP}}\right]\left(e^{-IRR_p^{IPP}t_2}-e^{-IRR_p^{IPP}T}\right)\right.$$

$$-\frac{r}{(T-t_1)\left(IRR_p^{IPP}\right)^2}\left[e^{-IRR_p^{IPP}t_2}\left(IRR_p^{IPP}t_2+1\right)-e^{-IRR_p^{IPP}T}\left(IRR_p^{IPP}T+1\right)\right]\Big\}\Big\}(1-p)(1-v_m)$$

$$=J\frac{\left(1+IRR_p^{IPP}\right)^{b+1}-1}{(b+1)IRR_p^{IPP}}\left(1+\frac{1-e^{-IRR_p^{IPP}T}}{T}\right)+J_M\Big[\left(1+\frac{1}{T-t_1}-\frac{t_1}{T-t_1}\right)e^{-IRR_p^{IPP}t_1}$$

$$-\left(1+\frac{1}{T-t_1}-\frac{T}{T-t_1}\right)e^{-IRR_p^{IPP}T}\Big].$$

$$(3.32)$$

The calculation of the IRR_p^{IPP} requires the application of a successive approximation method.

The formula for the Internal Rate of Return IRR is derived from (3.32) for $p=0$ and $v_m=0$, formula for IRR^{IPP} for $p=0$. The relative market value v_m for the interest rate IRR_p^{IPP} of the capital invested into purchase of a heat or power plant expected by the IPP for given investment expenditures J and J_M can be calculated from (3.32) [cf. formula (3.22)].

Moreover, the equation in (3.32) can be used to determine the market worth J of a power plant [compare formula (3.21)], which requires modernisation of the value given by J_M.

- Discounted Pay-Back Period of the investment $DPBP^{IPP}$

$$J_0\left[1+\frac{1}{T}-\left(1+\frac{1}{T}-\frac{t_1}{T}\right)e^{-rt_1}\right]+\Big\{N_{el}(1-\varepsilon_{el})t_A e_{el}^{t=0}\frac{1}{a_{el}-r}\left[e^{(a_{el}-r)t_1}-1\right]$$

$$-\frac{N_{el}t_A}{\eta_{el}}(1+x_{sw,m,was})e_{coal}^{t=0}\frac{1}{a_{coal}-r}\left[e^{(a_{coal}-r)t_1}-1\right]$$

$$-\frac{N_{el}t_A}{\eta_{el}}\rho_{CO_2}p_{CO_2}^{t=0}\frac{1}{a_{CO_2}-r}\left[e^{(a_{CO_2}-r)t_1}-1\right]-\frac{N_{el}t_A}{\eta_{el}}\rho_{CO}p_{CO}^{t=0}\frac{1}{a_{CO}-r}\left[e^{(a_{CO}-r)t_1}-1\right]$$

$$-\frac{N_{el}t_A}{\eta_{el}}\rho_{NO_X}p_{NO_X}^{t=0}\frac{1}{a_{NO_X}-r}\left[e^{(a_{NO_X}-r)t_1}-1\right]-\frac{N_{el}t_A}{\eta_{el}}\rho_{SO_2}p_{SO_2}^{t=0}\frac{1}{a_{SO_2}-r}\left[e^{(a_{SO_2}-r)t_1}-1\right]$$

$$-\frac{N_{el}t_A}{\eta_{el}}\rho_{dust}p_{dust}^{t=0}\frac{1}{a_{dust}-r}\left[e^{(a_{dust}-r)t_1}-1\right]-\frac{N_{el}t_A}{\eta_{el}}(1-u)\rho_{CO_2}e_{CO_2}^{t=0}\frac{1}{b_{CO_2}-r}\left[e^{(b_{CO_2}-r)t_1}-1\right]$$

$$-J(1+x_{sal,t,ins})\frac{\delta_{serv}}{r}(1-e^{-rt_1})-J_0\left[1+\frac{1}{T}-\left(1+\frac{1}{T}-\frac{t_1}{T}\right)e^{-rt_1}\right]\Big\}(1-p)(1-v_m)$$

$$+J_0\left[\left(1+\frac{1}{T}-\frac{t_1}{T}\right)e^{-rt_1}-\left(1+\frac{1}{T}-\frac{t_2}{T}\right)e^{-rt_2}\right]$$

$$+ J_M \left[\left(1 + \frac{1}{T - t_1} - \frac{t_1}{T - t_1} \right) e^{-rt_1} - \left(1 + \frac{1}{T - t_1} - \frac{t_2}{T - t_1} \right) e^{-rt_2} \right]$$

$$+ \left\{ N_{el}^M \left(1 - \varepsilon_{el}^M \right) t_A^M e_{el}^{M,t=t_1} \frac{1}{a_{el}^M - r} \left[e^{\left(a_{el}^M - r \right) t_2} - e^{\left(a_{el}^M - r \right) t_1} \right] \right.$$

$$- \frac{N_{el}^M t_A^M}{\eta_{el}^M} \left(1 + x_{sw,m,was} \right) e_{coal}^{M,t=t_1} \frac{1}{a_{coal}^M - r} \left[e^{\left(a_{coal}^M - r \right) t_2} - e^{\left(a_{coal}^M - r \right) t_1} \right]$$

$$- \frac{N_{el}^M t_A^M}{\eta_{el}^M} \rho_{CO_2} p_{CO_2}^{M,t=t_1} \frac{1}{a_{CO_2}^M - r} \left[e^{\left(a_{CO_2}^M - r \right) t_2} - e^{\left(a_{CO_2}^M - r \right) t_1} \right]$$

$$- \frac{N_{el}^M t_A^M}{\eta_{el}^M} \rho_{CO} p_{CO}^{M,t=t_1} \frac{1}{a_{CO}^M - r} \left[e^{\left(a_{CO}^M - r \right) t_2} - e^{\left(a_{CO}^M - r \right) t_1} \right]$$

$$- \frac{N_{el}^M t_A^M}{\eta_{el}^M} \rho_{NO_x} p_{NO_x}^{M,t=t_1} \frac{1}{a_{NO_x}^M - r} \left[e^{\left(a_{NO_x}^M - r \right) t_2} - e^{\left(a_{NO_x}^M - r \right) t_1} \right]$$

$$- \frac{N_{el}^M t_A^M}{\eta_{el}^M} \rho_{SO_2} p_{SO_2}^{M,t=t_1} \frac{1}{a_{SO_2}^M - r} \left[e^{\left(a_{SO_2}^M - r \right) t_2} - e^{\left(a_{SO_2}^M - r \right) t_1} \right]$$

$$- \frac{N_{el}^M t_A^M}{\eta_{el}^M} \rho_{dust} p_{dust}^{M,t=t_1} \frac{1}{a_{dust}^M - r} \left[e^{\left(a_{dust}^M - r \right) t_2} - e^{\left(a_{dust}^M - r \right) t_1} \right]$$

$$- \frac{N_{el}^M t_A^M}{\eta_{el}^M} \left(1 - u^M \right) \rho_{CO_2} e_{CO_2}^{M,t=t_1} \frac{1}{b_{CO_2}^M - r} \left[e^{\left(b_{CO_2}^M - r \right) t_2} - e^{\left(b_{CO_2}^M - r \right) t_1} \right]$$

$$- \frac{(J + J_M)(1 + x_{sal,t,ins}) \delta_{serv}^M}{r} \left(e^{-rt_1} - e^{-rt_2} \right)$$

$$- J_0 \left[\left(1 + \frac{1}{T} - \frac{t_1}{T} \right) e^{-rt_1} - \left(1 + \frac{1}{T} - \frac{t_2}{T} \right) e^{-rt_2} \right]$$

$$- J_M \left[\left. \left(1 + \frac{1}{T - t_1} - \frac{t_1}{T - t_1} \right) e^{-rt_1} - \left(1 + \frac{1}{T - t_1} - \frac{t_2}{T - t_1} \right) e^{-rt_2} \right] \right\} (1 - p)(1 - v_m)$$

$$+ J_0 \left[\left(1 + \frac{1}{T} - \frac{t_2}{T} \right) e^{-rt_2} - \left(1 + \frac{1}{T} - \frac{DPBP^{IPP}}{T} \right) e^{-rDPBP^{IPP}} \right]$$

$$+ J_M \left[\left(1 + \frac{1}{T - t_1} - \frac{t_2}{T - t_1} \right) e^{-rt_2} - \left(1 + \frac{1}{T - t_1} - \frac{DPBP^{IPP}}{T - t_1} \right) e^{-rDPBP^{IPP}} \right]$$

$$+ \left\{ N_{el}^{mod} \left(1 - \varepsilon_{el}^{mod}\right) t_A^{mod} e_{el}^{mod,t=t_2} \frac{1}{a_{el}^{mod} - r} \left[e^{\left(a_{el}^{mod}-r\right)DPBP^{IPP}} - e^{\left(a_{el}^{mod}-r\right)t_2} \right] \right.$$

$$- \frac{N_{el}^{mod} t_A^{mod}}{\eta_{el}^{mod}} \left(1 + x_{sw,m,was}\right) e_{coal}^{mod,t=t_2} \frac{1}{a_{coal}^{mod} - r} \left[e^{\left(a_{coal}^{mod}-r\right)DPBP^{IPP}} - e^{\left(a_{coal}^{mod}-r\right)t_2} \right]$$

$$- \frac{N_{el}^{mod} t_A^{mod}}{\eta_{el}^{mod}} \rho_{CO_2} p_{CO_2}^{mod,t=t_2} \frac{1}{a_{CO_2}^{mod} - r} \left[e^{\left(a_{CO_2}^{mod}-r\right)DPBP^{IPP}} - e^{\left(a_{CO_2}^{mod}-r\right)t_2} \right]$$

$$- \frac{N_{el}^{mod} t_A^{mod}}{\eta_{el}^{mod}} \rho_{CO} p_{CO}^{mod,t=t_2} \frac{1}{a_{CO}^{mod} - r} \left[e^{\left(a_{CO}^{mod}-r\right)DPBP^{IPP}} - e^{\left(a_{CO}^{mod}-r\right)t_2} \right]$$

$$- \frac{N_{el}^{mod} t_A^{mod}}{\eta_{el}^{mod}} \rho_{NO_x} p_{NO_x}^{mod,t=t_2} \frac{1}{a_{NO_x}^{mod} - r} \left[e^{\left(a_{NO_x}^{mod}-r\right)DPBP^{IPP}} - e^{\left(a_{NO_x}^{mod}-r\right)t_2} \right]$$

$$- \frac{N_{el}^{mod} t_A^{mod}}{\eta_{el}^{mod}} \rho_{SO_2} p_{SO_2}^{mod,t=t_2} \frac{1}{a_{SO_2}^{mod} - r} \left[e^{\left(a_{SO_2}^{mod}-r\right)DPBP^{IPP}} - e^{\left(a_{SO_2}^{mod}-r\right)t_2} \right]$$

$$- \frac{N_{el}^{mod} t_A^{mod}}{\eta_{el}^{mod}} \rho_{dust} p_{dust}^{mod,t=t_2} \frac{1}{a_{dust}^{mod} - r} \left[e^{\left(a_{dust}^{mod}-r\right)DPBP^{IPP}} - e^{\left(a_{dust}^{mod}-r\right)t_2} \right]$$

$$- \frac{N_{el}^{mod} t_A^{mod}}{\eta_{el}^{mod}} \left(1 - u^{mod}\right) \rho_{CO_2} e_{CO_2}^{mod,t=t_2} \frac{1}{b_{CO_2}^{mod} - r} \left[e^{\left(b_{CO_2}^{mod}-r\right)DPBP^{IPP}} - e^{\left(b_{CO_2}^{mod}-r\right)t_2} \right]$$

$$- \frac{(J + J_M)(1 + x_{sal,t,ins}) \delta_{serv}^{mod}}{r} \left(e^{-rt_2} - e^{-rDPBP^{IPP}} \right)$$

$$- J_0 \left[\left(1 + \frac{1}{T} - \frac{t_2}{T}\right) e^{-rt_2} - \left(1 + \frac{1}{T} - \frac{DPBP^{IPP}}{T}\right) e^{-rDPBP^{IPP}} \right]$$

$$\left. - J_M \left[\left(1 + \frac{1}{T - t_1} - \frac{t_2}{T - t_1}\right) e^{-rt_2} - \left(1 + \frac{1}{T - t_1} - \frac{DPBP^{IPP}}{T - t_1}\right) e^{-rDPBP^{IPP}} \right] \right\} (1 - p)(1 - v_m)$$

$$= J_0 \left(1 + \frac{1 - e^{-rT}}{T}\right) + J_M \left[\left(1 + \frac{1}{T - t_1} - \frac{t_1}{T - t_1}\right) e^{-rt_1} - \left(1 + \frac{1}{T - t_1} - \frac{T}{T - t_1}\right) e^{-rT} \right]$$

$$(3.33)$$

The calculation of the $DPBP^{IPP}$ value requires the successive approximation method.

The formula for the dynamic payback period $DPBP$ is derived from (3.33) for $v_m = 0$.

The introduction of investment expenditure J_M on modernisation of the heat and power plants into the mathematical models offer the statement of general form of them, and in the same manner, significantly increases their cognitive, and particularly, practical value. They enable the study and analysis of the influence of not only changes in price of energy carriers and specific cost associated with environmental charges over time, but also of the impact of the value of investment on the value of the market of heat and electricity as well as market worth of heat and power plants.

3.3 Conclusions

Mathematical models leading to statements regarding the value of heat and electricity market as well as market worth of power stations and heat and power plants presented here offer foundations for an extensive technical and economic analysis in issues arising in modern power engineering. The technical and economic analysis of energy market performed using these models lead to the statement of specific answers to the following questions (among others).

- Which particular energy technologies are worth investing? How much profit can be potentially derived from investing in the energy market?
- To what extent is this market affected by the price of energy carriers, specific tariff charges for polluting the environment, annual period of operation and electricity demand for internal load of the power station or heat and power plant, price of the CO_2 emission permission, investment needed on modernisation of the plants?
- What is the market price of power stations and heat and power plants?

In search of answers to these questions we need to remember that per capita energy consumption increases worldwide, and that its predicted increase will be significant. Demand is a measure of the size of the market and it mainly determines the value of a particular market.

The notation of the *NPV*, *IRR*, *DPBP* measures presented in this book and continuous mathematical models constructed by means of them has a number of tremendous advantages over discrete notation. It is noteworthy that the continuous notation offers the analysis of the mathematical models using differential calculus, without which it would be impossible or at least very difficult to follow relations showing how energy technology and energy efficiency of individual instruments as well as demand (annual production) for heat Q_A and electricity $E_{el,A}$ together with relations and scope of change in energy carriers price affect the output values of these models.

Reference

1. Bartnik, R., Bartnik, B.: Economic calculus in power industry, (Wydawnictwo Naukowo-Techniczne WNT), Warszawa (2014)

Index

© The Author(s) 2016
R. Bartnik et al., *Optimum Investment Strategy in the Power Industry*,
SpringerBriefs in Applied Sciences and Technology,
DOI 10.1007/978-3-319-31872-1

Printed in the United States
By Bookmasters